Maintenance Planning and Scheduling

Streamline Your Organization for a Lean Environment

Maintenance Planning and Scheduling

Streamline Your Organization for a Lean Environment

Timothy C. Kister
Bruce Hawkins

AMSTERDAM • BOSTON • HEIDELBERG • LONDON
NEW YORK • OXFORD • PARIS • SAN DIEGO
SAN FRANCISCO • SINGAPORE • SYDNEY • TOKYO

Butterworth-Heinemann is an imprint of Elsevier

Elsevier Butterworth–Heinemann
30 Corporate Drive, Suite 400, Burlington, MA 01803, USA
Linacre House, Jordan Hill, Oxford OX2 8DP, UK

⊗ Recognizing the importance of preserving what has been written, Elsevier prints its
books on acid-free paper whenever possible.

Library of Congress Cataloging-in-Publication Data
Kister, Timothy C.
   Maintenance planning and scheduling : streamline your organization for
a lean environment / Timothy C. Kister, Bruce Hawkins.
      p. cm.
   Includes bibliographical references and index.
   ISBN-13: 978-0-7506-7832-2 (alk. paper)
   ISBN-10: 0-7506-7832-1 (alk. paper)
   1. Plant maintenance—Planning. 2. Production management.
I. Hawkins, Bruce. II. Title.
   TS192 . K577 2006
   658.2′ 02—dc22   2005032928

British Library Cataloguing-in-Publication Data
A catalogue record for this book is available from the British Library.

ISBN-13: 978-0-7506-7832-2
ISBN-10: 0-7506-7832-1

For information on all Elsevier Butterworth–Heinemann publications
visit our Web site at www.books.elsevier.com

Printed and bound by CPI Group (UK) Ltd, Croydon, CR0 4YY
Transferred to digital print 2012

Table of Contents

    1.1 PURE REACTIVE, 3
    1.2 LIMITED PROACTIVE APPLICATIONS, 5
    1.3 BIRTH OF REAL MAINTENANCE, 6
    1.4 MILITARY TAKES THE LEAD IN MAINTENANCE, 10
    1.5 LONG JOURNEY TO LEAN THINKING, 12
        1.5.1 Lean Spills Over to Maintenance, 15
        1.5.2 Maintenance Operation Refinements, 16

    2.1 LEAN ORIGINS AND DEFINITIONS, 19
    2.2 LEAN ORGANIZATION: ELEMENTS
    AND PRACTICES, 22
        2.2.1 Current State, 25
        2.2.2 Future State, 26
    2.3 LEAN MAINTENANCE OPERATIONS, 28
        2.3.1 Fundamentals of Total Productive Maintenance, 29
            2.3.1.1 Work Execution, 33
            2.3.1.2 Planning and Scheduling, 34
            2.3.1.3 Reliability Engineering, 34
        2.3.2 Lean Refinements, 35
            2.3.2.1 Reliability Excellence, 35
            2.3.2.2 Teams and Teamwork, 36
            2.3.2.3 New Roles for Managers and Supervisors, 38
            2.3.2.4 Organizational Focus, 38
            2.3.2.5 Expanded Education and Training, 39
            2.3.2.6 Maintenance Optimization, 39
        2.3.3 Lean Maintenance: Prerequisite of the Lean
        Plant/Facility, 41

List of Figures

List of Tables

Acknowledgement

Very special thanks to Terry O'Hanlon who provided us with the insight into the importance of this unique topic: Lean Planning & Scheduling.

We also want to offer thanks to Jim Fei, Chairman and CEO of Life Cycle Engineering, Inc. Without Jim's understanding and commitment to the engineering and maintenance/reliability community, this book would not have been possible.

Acknowledgment

1

Historical View of Maintenance

In the greater scheme of things, maintenance has not been around very long. This was primarily because there was nothing to maintain for such a long, long time. A quick glance at Figure 1-1 will illustrate that relative to the time elapsed since the creation of Earth, the practice of maintenance has been with us for an infinitesimal amount of time. When starting from the beginning of the Paleozoic Era and equating time to a 12-hour clock, real maintenance has only been around for about 18 milliseconds. With such a short period of existence, it stands to reason that there are still many lessons to learn on the quest for perfection in the practice of maintenance.

Even though the first tools were used by hominids (human and human-like life forms; i.e., *Australopithecus, Homo erectus* and *Homo sapiens*), maintenance was slow to evolve. This is likely due to an extremely low level of intelligence in the earliest hominids. Figure 1-2 provides a depiction of *Australopithecus*—decide for yourself if you think he could embrace the concepts of planned, proactive preventive and predictive maintenance technologies. It was not until the coming of the metal ages—the Copper Age, Bronze Age and Iron Age, and following the evolution of the wheel as it progressed from your basic log to a solid disk and then into the spoked wheel (with metal hubs to fit onto an axle) that any true form of maintenance really became widespread. The timelines in Figure 1-1 trace the evolution of maintenance correlated to the evolution of Earth, human evolution and the evolution of technology.

The upper most time line—the Geological Time Line—represents the time from the formation of Earth, approximately five billion years ago, through the pre-Cambrian Era and into the Paleozoic Era, which consisted of six periods—from the Cambrian Period through the Permian Period. During the Paleozoic Era, life forms evolved from primitive marine invertebrates,

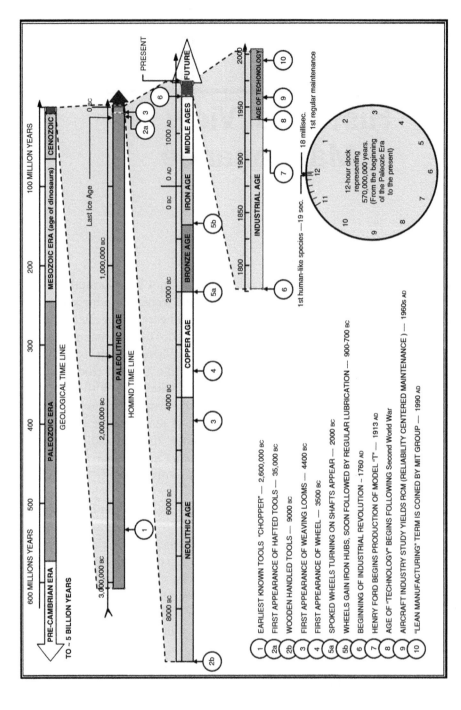

Figure 1-1 Historical Moments in Maintenance

Figure 1-2 Australopithecus

through initial vertebrate (e.g., Agnatha) life forms, fishes, amphibians and ending with the great expansion of reptiles. Following the Paleozoic Era was the Mesozoic Era, consisting of the Triassic, Jurassic and Cretaceous periods; dinosaurs first appeared during the Paleozoic Era and ultimately disappeared 167 million years later, marking the beginning of the Cenozoic Era. The Cenozoic Era consisted of the Tertiary Period, during which the land was dominated by mammals, birds and insects. It was also during the Tertiary Period that the earliest known hominid, or human life form made its first appearance as Australopithecus. The Quaternary Period, during which the Earth witnessed the rise of civilization and the first *Homo sapiens*, followed the Tertiary Period.

The next time line—the Hominid Time Line—represents, as per the theory of evolution, the evolving human forms from the earliest species of *Australopithecus* (Figure 1-2), during the time of the ancient (paleo) stone (lithic) age to the more modern forms of the new (neo) stone (lithic) age of *Homo erectus* and *Homo sapiens*. The lower two time lines break down the technological evolution of tools, machines and manufacturing and maintenance practices.

1.1 PURE REACTIVE

When tools were initially developed by human-like beings, the maintenance mode was "run to failure." The earliest tools, some discovered from as early as 2,600,000 BC, were "choppers" similar to that shown in Figure 1-3. Choppers were flat stones—river stones, which were chipped along a single face of one edge to form a rudimentary hacking tool. It is theorized that these choppers were used primarily for cutting through the skin and sinews of the animals that were hunted for food at that time and for digging. The chopper remained the only tool used by humanity for almost 2,000,000

Figure 1-3 Chopper Tool

years, until the appearance of the hand axe, a superior version of the chopper. In this tool, the entire surface of the rock was worked. Because both faces were chipped, the edge of the hand axe is termed a blade, which was considerably sharper than that of the earlier chopper. After 2 million years, the maintenance mode was still "run to failure."

Between 40,000 and 35,000 BC, the first appearance of "stone flake" tools, rather than choppers and blades, was seen. Flakes were retouched to make nosed scrapers, carinate (ridged) scrapers and end scrapers. Blades and burins, or chisels were made in several sizes using the punch technique. Bones and antlers were made into points and awls by splitting, sawing and smoothing split-base and bi-conical points, which provide evidence that hafting was also employed. An array of tools in use at the end of the Old Stone Age is shown in Figure 1-4.

In 9000 BC, when the new Stone Age began, axes, knives and other specialty tools were routinely fastened to wood or bone handles. At this point, the stone tools were re-flaked as the edges dulled and broken handles were repaired or replaced. Although this was still "run to failure," the concept of

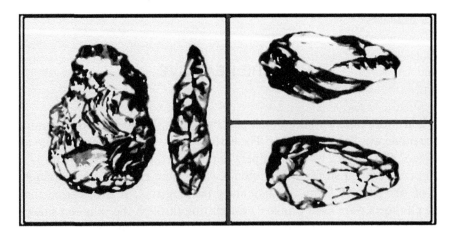

Figure 1-4 Stone Age Tools

"repair and overhaul" had made reuse of the most prized tools possible, which was an historical moment in maintenance.

Manufacturing on a larger scale than the making of personal tools began with the weaving of cloth. The weaving loom made its first appearance during the 5th millennium BC, around 4400 BC. These early looms consisted of bars or beams fixed in place to form a frame to hold a number of parallel threads in two sets, which together formed the warp. A block of wood was used to carry the filling strand through the warp. This fundamental operation of the loom remains unchanged although numerous improvements have been made. The weaving looms represented the first use of relatively standardized parts, and spares were pre-made so that broken parts of the looms could be replaced to avoid significant delays in the weaving (production) process. Maintenance was still "run to failure" but the art of maintenance now included the "maintenance stores" concept.

About 1000 years later, the wheel first appeared in Mesopotamia. A Sumerian pictograph, dating from about 3500 BC, depicts a conveyance that was equipped with wheels. The idea of wheeled transportation may have come from the use of logs for rollers, but the oldest known wheels were simple wooden disks consisting of three carved planks clamped together. Spoked wheels appeared about 2000 BC, when they were first used on chariots. Technological improvements followed, and with the onset of the Iron Age (roughly between 900 and 700 BC, depending on geographical location) included the use of iron hubs (centerpieces) turning on greased axles. Later the use of a tire in the form of an iron ring, which was expanded by heat and dropped over the rim and that, on cooling, shrank and drew the members tightly together. These improvements to the wheel brought about one of the more significant advances in maintenance as practiced until that time were regular, although unscheduled, preventive maintenance activities, indicating the first departure from the run to failure mode practiced for 2 million years. When the iron hub began to bind on the axle and slow the wheel's rotation, or when the hub–axle interface began to emit loud squeals and screeches, the wheel would be removed from the axle and grease would be reapplied to restore smooth, noise-free rotation. All maintenance practices to this point in history have shared the same characteristic that many maintenance operations still exhibit today—they were operating in a pure and untainted reactive maintenance mode.

1.2 LIMITED PROACTIVE APPLICATIONS

The first half of the Middle Ages are also referred to as the Dark Ages, although we now know that the Dark Ages were perhaps not as dismal as the name suggests. Many of the institutions created during this period profoundly influenced the character of civilization as it evolved in western Europe. The Christian Church was a significant institution of this type, but the Roman practices of law and administration also continued to exert an influence long after the departure of the Roman Legions from the western provinces of Europe.

Early waterwheels, used for lifting water from a lower to a higher level for irrigation, consisted of a number of pots tied to the rim of a wheel that was forced to rotate about a horizontal axis using running water or a treadmill. The lower pots were submerged and filled in the running stream; when they reached their highest position, they poured their contents into a trough that carried the water to the fields. The three power sources used in the Middle Ages—animal, water and wind—were all exploited by means of wheels. Horses were used to pull chariots and modified chariots or carts. Cattle and oxen often pulled the larger carts, or wagons. One method of driving millstones for grinding grain was to pull or push a long horizontal arm fitted to a vertical shaft connected to the stone with a horse or other beasts of burden. Waterwheels and windmills were also used to drive millstones for the grinding of grain into meal, gluten and flour.

The very first proactive maintenance practice was born out of the popularity of chariot races. Just before a race, chariot wheels were removed, hubs and axles were greased and the wheels reinstalled, even though the wheels were neither binding nor squealing. Chariot drivers anticipated the need for speed and knew that, besides a fast horse, well-greased wheels were essential.

1.3 BIRTH OF REAL MAINTENANCE

The preceding sections were a bit of history without much to do with the history of modern maintenance. It is offered as a little "tongue-in-cheek" treatise whose real purpose is to allow us all to realize that relative to the history of this planet we live on, our presence represents a mere tick of the clock. Hopefully, as this text continues, this view of ourselves will preclude taking ourselves too seriously, and open our minds to new concepts and ideas, allowing us collectively to rate a bullet on our own time line—the birth and evolution of the "Lean Maintenance Planner/Scheduler."

The Industrial Revolution, as it is referred to today, is generally acknowledged as beginning around 1750 and lasting until the First World War. The term Industrial Revolution, like many historical concepts, is a convenience to designate a period within which a preponderance of similar events occurred. There was no single event that triggered the Industrial Revolution, but there were sufficient innovations at the turn of the eighteenth and nineteenth centuries to justify this choice as one of the designated periods. A significant element of the Industrial Revolution was the advance in power technology. Before the beginning of this period, the major sources of power available to industry were four-legged energy and the power of wind and water. The use of steam power increased rapidly during the period and was the most significant energy delivery device, and remained so for most industrial purposes until well into the nineteenth century. Steam did not simply replace other sources of power; it transformed them. Steam motive power spurred the Industrial Revolution as more and more innovations of industry were developed using this medium.

During the period 1750 to 1830, the Industrial Revolution was largely confined to Britain. Keenly possessive of their head start, the British strictly prohibited the export of machinery, skilled workers and manufacturing techniques beyond their island empire. Their monopoly could not last forever, especially since some Britons envisioned highly profitable opportunities abroad. Continental European businessmen sought to lure British know-how to their countries and offered significant incentives. Two Englishmen, William Cockerill and John Cockerill, first brought the Industrial Revolution to continental Europe in Belgium by developing machine shops at Liège (*circa* 1807), and Belgium became the first country in continental Europe to be transformed economically by the Industrial Revolution. Like the British before them, the Belgian Revolution was oriented around iron, coal and textiles.

France was much slower to follow, and ultimately less completely industrialized than either Britain or Belgium. While Britain was establishing its industrial leadership, France was engaged in internal revolution and such an uncertain political environment discouraged large investments in industrial innovations. By 1848, although France had become an industrial power, it remained well behind Britain.

The first major impact on society brought about by the Industrial Revolution was the substitution of the factory for the domestic systemdue to the consequence of the mechanical innovations of that time. Four great inventions completely altered the manufacture of cotton goods—the spinning jenny, the waterframe, Crompton's mule (introduced in 1779) and the self-acting mule (first invented in 1792, but not brought into use until improvements were made to it in 1825). None of these innovations by themselves would have revolutionized the cotton industry, but in 1769 (the year in which Napoleon and Wellington were born), James Watt patented his steam engine. Sixteen years later, it was applied for the manufacture of cotton goods. The culmination of all of these inventions taken together marked the introduction of the factory system.

However, the invention with the greatest impact, and the one most fatal to domestic industry—the family run and operated enterprises—was the powerloom, patented in 1785, although it did not come into widespread use for several years. When the powerloom was first introduced, the workman was rarely injured. In fact, at first the new machinery raised the wages of spinners and weavers owing to the great prosperity it brought to the trade. In 15 years, cotton trade tripled itself. That period (1788–1803), has been called the "golden age" of cotton. A few years later however, the condition of the workman would become very different.

Meanwhile, the iron industry had been similarly revolutionized by the introduction of smelting by pit-mined coal brought into use between 1740 and 1750, and later followed by the application of the steam engine to blast furnaces in 1788. In the following eight years, the amount of iron being manufactured nearly doubled itself. With the rapid growth in production of both cotton goods and iron, it was inevitable that periods of overproduction would occur. As goods exceeded demand and stockpiles were created, the situation gave birth to the first major "maintenance style"—shutdown maintenance.

Manufacturing machinery of that time was subject to rapidly increasing failure rates with age. When failures occurred, production would be halted for repairs and some other manufacturer would steal away the merchants who had previously bought from the "down for repairs" producer. Although repairs were made as quickly as possible to resume production, the lost buyers often could not be regained. This resulted in manufacturers deliberately overproducing and stockpiling manufactured goods. When they had sufficient backlog of goods, these manufacturers would shut down their plant and repair/rebuild the production machinery. Often these shutdowns would last as long as one to two months. These planned shutdowns were the first real "planned maintenance" activities and continue to be practiced today, although to achieve entirely different objectives. In fact, the scheme of overproduction and shutdown maintenance, as originally conceived, continued for more than 100 years—well into the twentieth century.

At, and just after, the turn of the twentieth century, two milestone events dramatically changed manufacturing practices as well as the way in which maintenance was performed. The first of these was Henry Ford's creation of the modern assembly line and the second was First World War.

In 1860, 75% of the world's industrial production was concentrated in Western Europe, and of that, England accounted for about 50%. U.S. industrial production amounted to approximately 20% and the remaining 5% was spread over other parts of the world. In 1913, the year before the outbreak of First World War, the United States and Europe (including Great Britain) accounted for over 90% of the total industrial production of the world. The rest of the world accounted for only 10%. In the mid to late 1880s, U.S. industrial production went ahead of the British for the first time, and it has stayed well ahead, and also of all other industrial countries ever since (see Figure 1-5). Throughout the entire period that followed, Europe's percentage share of world industrial production steadily decreased, though absolutely it increased substantially. In the same period, the U.S. share steadily increased both relatively and absolutely. Among those responsible for this increase was Frederick W. Taylor, who in 1881 began delving into the organization of manufacturing operations at the Midvale Steel Company. Taylor refined both the methods and the tools used in the various stages of steel manufacture, permitting workers to produce significantly more with less effort. However, the individual most responsible for the dramatic productivity increase in the United States was Henry Ford.

The Ford Motor Company was incorporated in 1903 and built several different automobile models between 1903 and 1908. In 1908, Ford introduced the "Model T" and shortly thereafter announced that the Model T would be the only automobile that he would manufacture. In 1908, Ford produced 100 automobiles a day. The time required to assemble each Model T was 728 minutes. Henry Ford's assembly line underwent some serious fine-tuning

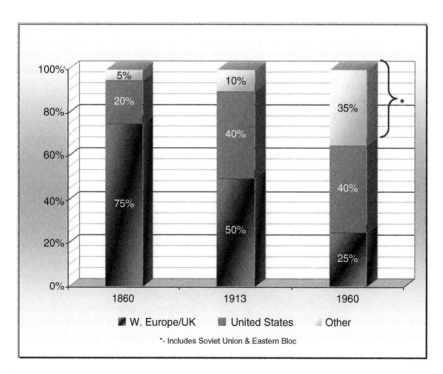

Figure 1-5 Distribution of World Industrial Production

between 1908 and 1913. Assembly time went from 728 minutes to 93 minutes while production simultaneously increased from 100 automobiles per day to nearly 1000. For Ford to maintain this production rate, assembly line stoppages could not be tolerated. One of the primary methods to ensure the smooth and consistent flow of each chassis through the assembly line was "planned maintenance."

1.4 MILITARY TAKES THE LEAD IN MAINTENANCE

The second event to have a profound influence on maintenance operations was the First World War. At the war's outbreak in 1914, troop mobility quickly became a distant memory as trench warfare took hold along entire fronts. Army commanders of the time defined the role of the aircraft, invented just 10 years earlier, as one of reconnaissance. Scout planes were used to pinpoint enemy trench locations for artillery spotting. Unfortunately, the artillery of the time (both guns and shells) were something less than consistent when it came to hitting small targets, small as in trenches. In frustration, pilots began to take matters into their own hands. What began as pilots shooting at their enemy counterparts with pistols and rifles, evolved quickly as pilots realized that they could also inflict significant damage within the trenches using fuselage-mounted machine guns. Some pilots even dropped bombs by hand, on the entrenched troops (Figure 1-6).

The evolving role of aircraft gave rise to greatly increased numbers of aircraft being employed just behind major fronts. It also gave rise to more missions that meant more hours in the air, which in turn gave rise to more

Figure 1-6 First World War Sopwith Camel

equipment failures. In general, major equipment failures in all aircrafts were most often fatal. In the warfare scenario, the fatality rate went even higher. Pilots, who were keenly interested in returning from each mission, developed pre-mission checklists for determining a plane's combat airworthiness. By the end of the First World War, the pilots' checklists had evolved into a maintenance check-off sheet. Over time and from the experiences of hundreds of pilots, the maintenance check-off sheets became thorough and comprehensive documents for identifying required maintenance and repairs. Their use became so common that following the end of the War, they became standard post- and pre-flight checklists for all aircraft operations, both military and nonmilitary.

Another development of the First World War was the armored tank. The concept for the tank was proposed by an English Army engineer officer assigned to observe and report on the European theater (civilian correspondents of the time were not permitted in combat areas). The engineer officer-turned-correspondent witnessed many frontal assaults on entrenched German Army positions in which the Allied soldiers first met, and then became entangled in barbed wire and subsequently received devastating machine-gun fire that literally wiped out thousands during just one assault. Shortly after witnessing one of those hapless assaults, the engineer officer observed a French farmer at work. He was driving a field tractor, carefully and successfully navigating his way through vacated battlefronts complete with barbed wire, trench parapets and protective berms. The engineer officer immediately envisioned the tractor equipped with a protective box of armor plate, treads and machine guns making the previously observed assault on the entrenched Germans. He knew he had more than just a good idea.

Winston Churchill, the British First Lord of the Admiralty during the First World War, a position equivalent to the U.S. Secretary of the Navy, was just beginning to address the British armed forces' problem of overcoming the combination of barbed wire and machine gun. Why the navy, you ask? As part of its responsibility for protecting Britain against aerial attack by the Germans, the British Navy had established air bases at Dunkirk on the French coast. Armored-car squadrons equipped with vehicles that were not very satisfactory for crossing ditches and obstacles that got in their way protected these bases. When Churchill read the engineer officer's memorandum describing the armored tractor, he knew it could be a significant weapon. Churchill championed the development of the armored tank, which he termed the "landship" and which was initially built around a Holt Caterpillar Tractor, manufactured by an American Company, and Willie, the first Allied First World War tank as it came to be known, took to the front (Figure 1-7).

Figure 1-7 Mark I "Willie" (WWI Tank). *Weight*: 27-tons, *Dimensions*: 6.4′ × 14.3′ × 8.0′, *Armor* (max): 0.47″, *Range*: 23 miles, *Speed* (max): 4 mph, *Weapons*: 5–7.7 mm Machine Gun, Crew: 8

As with aircraft, Willie's reliability was a life and death issue. Although Willie could easily overrun enemy positions, immobility rendered the tank a death trap. Maintenance not only became an important issue for the tank, but it actually underwent some sophisticated evolution. Engine oil was replaced on a time-based schedule; other engine maintenancelike oil, fuel and air filter replacements, etc.,as well as tread replacement became condition-based maintenance actions. In addition, weaponry (e g. the machine gun)was completely torn down, cleaned and reassembled daily when in combat. All of these innovations in maintenance originated with the military, due to the ultimate nature of the finality of major equipment failure.

This driving force, that is the life and death nature of equipment failure, gradually took second place as political forces took hold and slowly began to create the military industrial complex. The driving force of having a "military machine second to none" soon became the "innovation enabling" mechanism in the field of maintenance, and in fact, in nearly all technological advancements being made. However, one thing that the military establishment was not especially interested in was "Lean Maintenance." Until only recently the military approached maintenance as they did warfare— throw large enough sums of money and people at a given problem and a solution was sure to be found.

1.5 LONG JOURNEY TO LEAN THINKING

Immediately following the First World War, few changes were made in the ways in which maintenance was planned and performed. The use of the mass production assembly line, as Henry Ford had drastically refined it, rapidly spread throughout American manufacturing. However, there was a brief depression in the United States in 1920 and 1921 that was quickly forgotten by the public at large due to the burgeoning economy brought about by the spread of efficient mass production manufacturing techniques and the lower prices that resulted. However, the depression period did result in one lasting effect. It ended Ford's solitary domination of the automobile industry and led to the joint dominance of the "big three" (Ford, General Motors and Chrysler Corp.). The sharing of the American automobile market provided increasing efficiencies to manufacturing processes through the new-found competition among the big three.

The stock market crash in 1929 and the major depression that followed, marked a turning point in many facets of American life, not the least of which was in the make-up of the manufacturing industry. At the same time, it would be inaccurate to say that the depression radically altered the course that the industry was following. It did not. The pattern of development was clearly established by 1929; the crash merely confirmed the trend toward concentration in the larger manufacturing entities and expedited that process by delivering a crushing blow to many of the small producers who had been struggling desperately for life even in prosperous times.

For the survivors, the depression necessitated some readjustment in their thinking, from an optimistic belief that the demand for more and more manufactured goods was insatiable to a realization that they were facing a scaled down market with demand well below their own capacity to produce. Even though the number of manufacturers had been sharply diminished, competition between them was intensified. One result was a greater emphasis on the marketing side of the business, but not at the expense of technological development. On the contrary, there was more incentive than ever to keep offering a better product for less money. Table 1-1, which shows the prices for new automobiles at various times, is typical of price trends for nearly all consumer goods. In terms of consumer gains in the array of goods available, as well as in quality, product innovation and the economics of competitive pricing – the decade of the 1930s was well ahead of the 1920s.

However, precisely because of this concentration of manufacturing in fewer, but larger and better financed plants and because of the additional production capacity created that went well beyond demand, there was little incentive for innovations in maintenance. Breakdowns in production lines, while troublesome, posed no real threat to either profits or sales. Many plants operated with just one or two full-time maintenance personnel. Manufacturing plants would bring in temporary mechanics, pipe fitters and electricians as required, to deal with unexpected failures. Similarly, during planned maintenance shutdowns, the maintenance staff would temporarily grow to meet the scheduled outage workload and shrink back down in size when the work is completed. It was not an especially good time to be specialized in maintenance, at least not if steady work was desired.

Table 1-1
Average F.O.B. Retail Price of New U.S. Manufactured Automobiles

Year	Average Price (US$)
1899	1559
1909	1719
1919	1157
1929	828
1939	845
1947	1580

Then, as is wont in times of depressed economies, along came the Second World War to help turn things around. Everything changed dramatically during the Second World War due to significant manufacturing manpower shortages and an ever-increasing demand on production. The technology of manufacturing was forced to develop more mechanization to offset the labor shortages and to meet the growing demand for War materials as well as consumer goods. Although there has been much talk about the shortages of goods during the War, they were primarily isolated to foreign-procured goods (e.g., sugar) and to industry sectors that had converted to war materials production (e.g., automobile manufacturers). However, as technological advances in weaponry manufacture were made, enterprising individuals applied many of the new technologies to the manufacture of consumer products. The war also produced a brand-new phenomenon—the working woman, which ultimately led to more disposable income per family and a higher standard of living at all levels of society.

The combined impact of these events had another incremental impact on maintenance. Suddenly, the reliability of equipment became important and production downtime became a concern to everyone. The newfound stature of the maintenance mechanic (millwright) allowed the maintenance organization to develop and implement planned, periodic, preventive maintenance programs; something that they had long known was needed. Unleashed to pursue the refinement of their trade, maintenance workers had made a gigantic leap that effectively took maintenance out of the "reactive only" mode of operation. Many more innovations were to come in the next 20 to 30 years.

While this leap into planned, periodic, preventive maintenance was being endorsed by more and more manufacturers in the Western world, Japan was trying desperately to recover from the devastation that the war had brought on their economy and their manufacturing sector. One company in particular that had been operated by the Toyoda family was forced to scrap their previous plant operations and completely reorganize because of their support of the Japanese war machine. The company was renamed Toyota. At the Toyota Automobile Group, Taiichi Ohno; the manager of machining operations attempted to create a viable operation in spite of severe material shortages brought about by the war. Gradually, Ohno improved existing processes to better support assembly operations. For the systems that were developed (leading ultimately to the Toyota Production System or TPS), Ohno is credited with two concepts adapted from U.S. practices. The first concept was the assembly line production system, which Ohno derived from Henry Ford's book *Today and Tomorrow* first published in 1926. The second concept was the supermarket operating system used in the United States, a system that Ohno observed during a visit in 1956. The supermarket concept provided the basis of a continuous supply of materials just as the supermarket provides a continuous supply of goods to the (shelves) consumer.

More than 30 years later, Ohno's Toyota Production System would be credited bymany as the birth of "Lean Manufacturing." Interestingly, the term Lean Manufacturing did not originate in Japan. It was originally coined in the book *The Machine That Changed the World*, published in 1990. In 1996 a second book, *Lean Thinking*, was published that defined most of the terms and practices employed in today's Lean Enterprise, which came first to manufacturing and second to manufacturing plant maintenance operations (Chapter 2 will define many of the terms and practices associated with "Lean Operations," whether in the manufacturing sector or any other enterprise.) Both books were authored by an MIT research group studying, among other things, the Toyota Production System (TPS), whose leaders were James Womack and Daniel Jones, of the US and UK, respectively—and not by Ohno or anyone directly associated with the TPS.

From the seeds sown during the first industrial revolution, beginning roughly around 1750, to the last decade of the twentieth century, a journey of more than 240 years, grew Lean Thinking, which led to Lean Manufacturing and then to Lean Maintenance.

1.5.1 Lean Spills Over to Maintenance

An initial concept of Lean Manufacturing was that it needed to only involve the production department, and if organized as independent of production, the purchasing/supply (raw materials and outside parts/assemblies that were input to production) departments. The most basic or fundamental premise of Lean Thinking is the identification and elimination of waste. Lean manufacturers jumped directly into this process of eliminating waste in their production processes and many were quite successful. Unfortunately, that very success was the root of their ultimate failure. Manufacturers quickly discovered that, as waste is eliminated from the production process, and available equipment operating time increases, reliability issues also begin to increase. Maintenance operations were not equipped to provide production equipment reliability levels needed to keep pace with the lean production teams. Perhaps they had gotten the proverbial cart before the (just as proverbial) horse. Maintenance was about to make another giant stride forward in the stature with which they were viewed by management.

Although even today it is not universally accepted as a law of Lean Manufacturing, the fact is that implementing Lean Maintenance in a manufacturing plant is prerequisite to implementing actual Lean Manufacturing practices. If those holdouts would examine three of the fundamental laws of

---

FUNDAMENTAL LAWS OF MANUFACTURING MAINTENANCE

- Properly maintained manufacturing equipment makes many, quality products
- Improperly maintained manufacturing equipment makes fewer products of questionable quality
- Inoperable equipment makes no products.

---

manufacturing maintenance, they might more readily recognize this prerequisite condition.

A manufacturing plant with intentions of implementing Lean Manufacturing should begin with a few essential preparations. One of the most important preparations is the configuration of the maintenance organization to facilitate Lean Maintenance first and Lean Manufacturing next.

1.5.2 Maintenance Operation Refinements

Maintenance Operation processes must be refined in order to support the Lean Plant Operation, while, at the same time supporting the maintenance operation's own Lean makeover. The number one objective for all maintenance organizations everywhere is the maintenance of equipment reliability. Additional objectives for Lean Maintenance Processes include:

- plan and schedule the maintenance workload to:
  - o maintain the work backlog within prescribed limits by providing for forecasted level resource requirements;
  - o create achievable daily schedules.
- continually reduce equipment downtime and increase availability through the establishment of a preventive/predictive maintenance program (including failure analysis) that is designed, directed, monitored and continually enhanced by maintenance engineering;
- ensure that work is performed efficiently through organized planning, level scheduling, optimized material support and coordinated work execution;
- establish maintenance processes, procedures and best practices to achieve optimal response to emergency and urgent conditions;
- create and maintain measurements of maintenance performance and effectiveness;
- create and provide meaningful management reports to enhance control of maintenance operations;
- provide quality, responsive maintenance service in support of operational need.

In addition to process refinements, successful Lean Maintenance Operations also require organizational refinements. In refining organizational relationships, there are two critical characteristics that need to be achieved.

1. Maintenance management should be structured level with production management.
2. Maintenance is seen as a supportive service to production vs. a subordinate responder.

There should always be a current and complete organizational chart that clearly defines all maintenance department reporting and control relationships, as well as relationships with other departments. The organization should clearly show responsibility for the three basic maintenance responses: (i) routine, (ii) emergency and (iii) backlog relief.

Interfaces between production and maintenance should be clear and divisions between roles, responsibilities and authorities should be well defined within the organizational structures. The maintenance organizational structure should recognize three distinct (separate but mutually supportive) functions, so that each basic maintenance function receives the primary attention required:

Work Execution
Planning and Scheduling
Maintenance Engineering

It has taken an exceedingly long time for Maintenance Operations to reach this point. Real advancements in the art of maintenance have only been realized during the last 240 years and 95% of these advancements have occurred during the last 25 to 35 years. Having come this far, this fast, makes it incumbent on today's generation of maintenance professionals to continue to aggressively push forward with improvements in the practice of maintenance until every plant and facility is reaping the benefits of maximum equipment reliability.

2

The Plant/Facility Lean Environment

The primary focus of this book is placed on maintenance planning and maintenance scheduling as performed in the Lean Environment. The qualifier in the lean environment introduces a host of new considerations for traditional practices. These considerations are new because these were not involved in traditional, or pre-Lean Maintenance planning and maintenance scheduling activities. Although all planning and scheduling activities will be thoroughly developed during the course of this book, it is nonetheless important to provide a comprehensive description of the Lean Environment as it is differentiated from the pre-Lean Environment.

2.1 LEAN ORIGINS AND DEFINITIONS

Chapter 1 briefly touched on the origins of "Lean Manufacturing." Taiichi Ohno is generally acknowledged as the father of Lean Manufacturing, and therefore with Lean Thinking. Ohno's Toyota Production System (TPS)—a Lean Manufacturing system—although created by Ohno, was conceived by Henry Ford in the first decade of the twentieth century and subsequently refined during the second decade. When Ford brought out his "Model T" in 1908, he introduced the first, efficient assembly line production process to the world of manufacturing. The assembly line employed the precise timing of a constantly moving conveyance of parts, subassemblies and assemblies, ultimately culminating in the creation of a completed Model T chassis. As a completed chassis rolled off the assembly line, 10 to 15 more were located at various stations along the assembly line, gradually being built-up eventually to roll out as another completed Model T.

Five years of fine-tuning the various operations, eliminating the wasted time in each of them and adjusting the rate at which parts and assemblies were being input to assembly operations eventually reduced the initial Model T assembly time of 728 minutes in 1908 to 93 minutes in 1913. This was manufacturing's introduction to Lean Thinking.

Sakichi Toyoda was the founder and owner of Toyoda Loom Works, Japan's largest loom manufacturing operation. In 1936, Sakichi Toyoda expanded his operations to include an automobile manufacturing group. Sakichi Toyoda appointed his son, Kiichiro Toyoda, as managing director of the new operation. Kiichiro Toyoda traveled to the Ford Motor Company in Detroit to spend a year studying the American automotive industry and returned to Japan with a thorough knowledge of the Ford production system. Kiichiro set about to not only adapt the system to smaller production quantities, but also to improve on the basic practices. In addition to the smaller production quantities, Kiichiro's system more precisely managed the logistics of materials input to coincide with production consumption. Kiichiro developed an entire network of suppliers capable of supplying component materials when needed. Within the Toyoda Group, the system was referred to as just-in-time (JIT).

Following the Second World War, the restructured Japanese government forced the Toyoda Group to reorganize in 1950. Kiichiro resigned and his cousin, Eiji Toyoda was named the new managing director. Like Kiichiro, Eiji also went to the United States to study the American system of automobile manufacturing.

At the reconstructed Toyoda Group Automotive Operations, renamed as the Toyota Automobile Group, Taiichi Ohno (see Chapter 1) managed the machining operations. His development of improved methods for supporting the automotive assembly operations was largely responsible for the success of the restructured company. Together with the implementation of quality initiatives, provided by Shigeo Shingo who was a quality consultant hired by Toyota, and the incorporation of statistical process control methods brought to Japan by Dr. W. Edwards Deming, the processes defined the TPS. The TPS quickly became not only the most successful manufacturing operation in Japan, but also embodies all of the present-day attributes of Lean Manufacturing.

Lean, and Lean Thinking, can most simply be described as the elimination of waste and creation of value for the customer (which derives from elimination of waste). There are seven categories of waste (the Seven Deadly Wastes) according to the theory behind Lean Thinking.

1. Overproduction—excess production and early production;
2. Waiting—delays—poor balance of work;

3. Transportation—long moves, re-distributing, pick-up/put-down;
4. Processing—poor process design;
5. Inventory—too much material, excess storage space required;
6. Motion—walking to get parts, tools, etc., lost motion due to poor equipment access;
7. Defects—part defects, shelf life expiration, process errors, etc.

The complexity of Lean, if there is any, is in identifying waste and then eliminating it. Over the course of the last four decades, the processes employed in the TPS have been refined, provided catchy names and defined within neat little boxes. The term "Lean" applied by James Womack et al. (see Chapter 1) was readily adopted by manufacturers, who preferred it to "Toyota Production System" for defining their special manufacturing style. The following concepts apply to Lean Manufacturing Practices.

Support for waste elimination through

- customer focus;
- committed management;
- doing it right the first time (quality control);
- enhanced customer value (from quality and price control);
- just-in-time (JIT) systems;
- integrated supply chain (from JIT);
- making and sustaining cultural change (personnel attitudes and ways of thinking);
- measurement (Lean Performance) systems;
- optimized equipment reliability;
- plant-wide lines of communication;
- standardized work practices;
- value-creating organization;
- winning employee commitment/empowering employees.

Take special note of the last item or concept in this list. Many people believe (erroneously) that the objective of Lean Operations is to reduce the size of the work force. It would be difficult indeed to gain employee commitment for a program whose intent was to eliminate their jobs. Instead, Lean empowers employees by endowing them with real decision-making authority within their particular processes. Although reducing the amount of labor required is an objective—and a natural byproduct from improved efficiencies and waste elimination—of Lean practices, reducing the size of the work force *is not* an objective of Lean Thinking. Thus, it introduces another objective to define new roles (roles that add value to the total manufacturing process), for many employees, and allowing normal attrition to account for the reduced labor requirements. Reducing waste is the preeminent objective in the Lean Operation and it must become everyone's responsibility even though difficult to enforce responsibility when employees are in fear of losing their job.

What is Lean Manufacturing? Here is one definition.

> *Lean Manufacturing is the practice of eliminating waste in every area of production including customer relations (sales, delivery, billing, service and product satisfaction), product design, supplier networks, production flow, maintenance, engineering, quality assurance and factory management. Its goal is to utilize less human effort, less inventory, and less time to respond to customer demand, less time to develop products, and less space to produce top quality products in the most efficient and economical manner possible.*

2.2 LEAN ORGANIZATION: ELEMENTS AND PRACTICES

It is also helpful to understand the Lean Environment through the five standard practices for the application of Lean Thinking:

1. *Specify* what does and does not create value from the customer's perspective and not from the perspective of individual firms, functions and departments.
2. *Identify* all the steps necessary to design, order and produce the product across the whole value stream to highlight nonvalue-adding waste.
3. Make those actions that create value *flow* without interruption, detours, backflows, waiting or scrap.
4. Only make what is *pulled* by the customer.
5. Strive for perfection through *continuous improvement*; removing successive layers of waste as they are uncovered.

These standard practices are fundamental to the identification and elimination of waste. They are easy to remember (although not always easy to achieve) and should be the guide for everyone in the Lean Organization.

Some of the more common techniques employed in the Lean Manufacturing or Process Plant are delineated in Table 2-1. They are grouped by general plant objective categories, although there is considerable latitude in how and where the techniques can be used. Many of the terms in this table will be new or foreign (many are foreign) to you, but you will come to understand them quite well. The "how and where" to apply these techniques must be defined by a thorough and probing analysis of the total manufacturing operation. Value stream mapping and process mapping are useful tools for this current state analysis. All too often, the next step in applying Lean Thinking is sidestepped and the techniques in Table 2-1 are immediately exercised. However, before using these techniques, the *future state* must be defined. The future state is a complete definition of what the operation is to become through the application of Lean Thinking. A journey's path is elusive unless both the starting and ending points are known in advance.

Most companies rely on production volume as their ultimate test for success. Lean is not about productivity, and this is hard for many manufacturers to accept. It is about removing waste from the manufacturing process and building quality. In the Lean Organization scheme, all of the business processes of the plant have common goals, i.e., gaining the competitive advantage. Lean means timely billing just as much as it means skilled, accurate machining. It means efficient sales and advertising just as much as it means reliable production equipment. All departments working together to relay and share information and data, to identify and correct problems, to maintain safe workspaces on the shop floor as well as in the business offices are fundamental elements of the lean operation.

Lean is a comprehensive package that includes reducing inventory, standardizing work routines, improving processes, empowering workers to make decisions about quality and waste elimination, soliciting worker ideas, proofing for mistakes, applying just-in-time delivery and using a Lean supply chain.

Lean is about waste reduction and customer focus. It is also about quality the first time and continuous improvement and also about problem solving. However, perhaps above all, Lean Thinking is about people. Unlike traditional manufacturing organizations, people are not the problem in the lean enterprise; people are the problem solvers.

Who knows more about problems with a step in the manufacturing process, the shop-floor operator or the middle level manager in his office filling out forms? Moreover, who is more likely to know what the solution is to the problem with a step in the manufacturing process? Lean empowers the shop-floor operator, encourages his involvement in waste reduction, customer relations and product quality, and in continuous improvement in all facets of the Lean enterprise.

Table 2-1
Manufacturing's Lean Processes

Waste Identification and Elimination	Quality Control
Seven Deadly Wastes	Statistical Process Control (Cp, Cpk)
Project Focus (One To Three Months)	Analysis of Variance (ANOVA)
Kaizen Events (One Week)	Design of Experiments, Taguchi
DMAIC–Design/Measure, Analyze, Improve, Control	Response Surface Methodology (RSM)
Value-Stream Mapping	Regression Analysis
Workplace Organization	Jidoka (Error Proofing)
Standardized Work	Six Sigma Techniques and Principles
5-S	Variation Reduction
Visual Management	**Inventory Reduction**
Measurement System Assessment (Gage R&R)	Just-in-Time (JIT) Vendor Selection
Yellow, Green, Black and Master Black Belts	Pull System (Kanbans)
Production Capacity	**Equipment Reliability**
Production Smoothing	Total Productive Maintenance (TPM)
One Piece Flow (Takt Time)	Autonomous Maintenance
Evolutionary Operation (EVOP)	Maintenance Engineering
Process Stability	FMEA (Failure Modes and Effects Analysis)
Set-Up Time Reduction (SMED)	Root Cause Analysis and Hypothesis Tests
Cellular Manufacturing	Team-Based, Multi-Skilled Workforce
Balanced Work Flow	

2.2.1 Current State

In order to eliminate waste, one must know where it exists. Therefore, the first step in identifying waste in any process is to define the *current state* of the process, which is done by beginning at the end of the process and mapping each step of the process by following it in reverse. In a production plant, for example, the process of a single production line would end as the finished product rolls off the line for quality assurance (QA) inspection and packaging. Each preceding step to "rolling off the line" is mapped to define the current state of the process. In the Lean Environment, the most commonly used method for mapping the current state of a process is referred to as *value stream mapping*.

Value Stream Mapping is a powerful tool for "seeing" a process, identifying the nonvalue-adding components and re-creating the process as a *value stream*. The mapping process employs several standard map symbols that were created for manufacturing processes. They are useable for the maintenance operation as well, but the important thing in value stream mapping is that the map be easily understood, so if you are more comfortable using symbols that you devise, use them. Some of the more common symbols are shown in Figure 2-1.

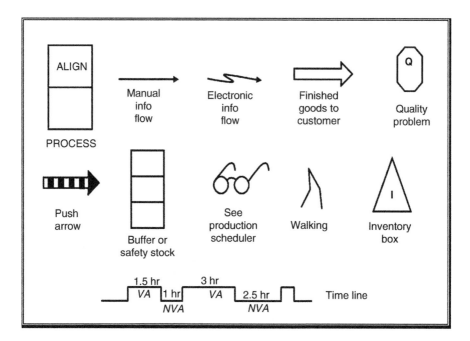

Figure 2-1 Symbols Used in Value Stream Mapping

Another often-used process mapping technique dating back to the early 1900s is the original system developed by Frank Gilbreth, which is still very useful. The Gilbreth approach is highly visual and discriminates between waste and value-added activity very clearly. It is also simple and easily used by even untrained groups. The symbology of this technique is illustrated and defined in Figure 2-2. When using this method of process mapping, it is necessary to provide annotation of each symbol by describing the event as concisely as possible and indicating the time required. Appendix C contains an example of this mapping technique. Appendix C also provides additional value stream mapping symbols as well as an example of a value stream process map.

2.2.2 Future State

Whatever process mapping system you decide to use, its application is the same. It employs a six-step waste identification and elimination process illustrated in Figure 2-3 and is outlined below.

1. Select the process for assessment, and starting at its end point, carefully map out each stage/activity of the process.
2. Analyze the process map by examining each map symbol and attempt to "drill-down" to additional process steps within each mapped step. Continue until the team agrees that all steps of the process have been mapped. This results in the *current state* map.
3. Re-analyze the *current state* map to identify all *nonvalue-adding* activities.
    (a) Remove the nonvalue-adding activities or develop value-adding "work-arounds" and remap the process.
    (b) Re-analyze the new map for workability and additional nonvalue-adding activities. Continue until the entire team agrees that the process is now workable and consists of only value added activities and "impossible to remove" nonvalue-adding activities.
4. The resulting map constitutes the definition of the process' *future state*.
    (a) Create a listing of all the actions needed to remove the nonvalue-adding activities as well as any value added work-arounds developed.
    (b) The listing of the actions needed to remove the nonvalue-adding activities as well as any value added work-arounds developed constitutes the steps of an *action plan* for modifying (removing the waste) the selected process.
5. Write-up an action plan that will be applied to the process' current state to move it to the (value added) future state. Submit the action

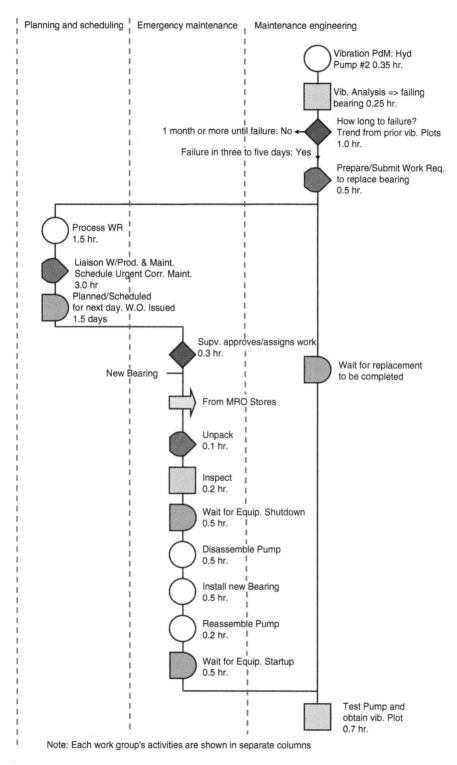

Planning and scheduling ¦ **Emergency maintenance** ¦ **Maintenance engineering**

Vibration PdM: Hyd Pump #2 0.35 hr.

Vib. Analysis => failing bearing 0.25 hr.

1 month or more until failure: No ◄ How long to failure? Trend from prior vib. Plots 1.0 hr.

Failure in three to five days: Yes

Prepare/Submit Work Req. to replace bearing 0.5 hr.

Process WR 1.5 hr.

Liaison W/Prod. & Maint. Schedule Urgent Corr. Maint. 3.0 hr

Planned/Scheduled for next day. W.O. Issued 1.5 days

Supv. approves/assigns work 0.3 hr.

Wait for replacement to be completed

New Bearing

From MRO Stores

Unpack 0.1 hr.

Inspect 0.2 hr.

Wait for Equip. Shutdown 0.5 hr.

Disassemble Pump 0.5 hr.

Install new Bearing 0.5 hr.

Reassemble Pump 0.2 hr.

Wait for Equip. Startup 0.5 hr.

Test Pump and obtain vib. Plot 0.7 hr.

Note: Each work group's activities are shown in separate columns

Figure 2-2 Gilbreth Process Mapping Symbology

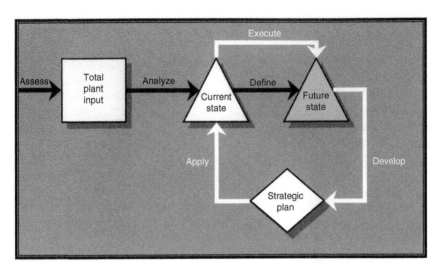

Figure 2-3 Implementing Lean Practices with Process Mapping

plan together with both the current state and future state process maps, for approval and authorization.

6. Execute the process' action plan in accordance with approval guidelines.

2.3 LEAN MAINTENANCE OPERATIONS

As pointed out in Chapter 1, a primary requirement for a Lean Manufacturing Plant is (production) equipment reliability. Maintenance practices, in turn, are a primary determinant of equipment reliability. The reactive style of maintenance practiced during the greatest part of the history of maintenance must be replaced with proactive *([pro- + reactive]: acting in anticipation of future problems, needs, or changes)*[1] maintenance practices in order to achieve the levels of equipment reliability necessary to sustain lean manufacturing goals and objectives. Do you recall the "Fundamental Laws of Manufacturing Maintenance" from Chapter 1? Well, there is a fundamental maintenance program that is the very foundation of lean maintenance and it's appropriately termed *Total Productive Maintenance* or *TPM*. Note the word "Productive." TPM is formulated to achieve maximum equipment reliability in support of production.

---

[1] *Merriam-Webster's 11th Collegiate Dictionary.*

2.3.1 Fundamentals of Total Productive Maintenance

TPM objectives include the elimination of all accidents, defects and breakdowns. TPM is team-based, proactive maintenance that involves every level and function in the organization, from top executives to the shop floor. TPM addresses the entire production system life cycle and builds a solid, shop floor-based system to prevent all losses. TPM activities should focus on results. One of the fundamental measures of performance used in TPM is Overall Equipment Effectiveness or OEE.

$$OEE = (\text{Equipment Availability}) \times (\text{Performance Efficiency}) \times (\text{Rate of Quality})$$

World-class levels of OEE start at 85% based on the following values:

$$90\% \text{ (Equipment Availability)} \times 95\% \text{ (Performance Efficiency)} \times 99\% \text{ (Rate of Quality)} = 84.6\% \text{ OEE}$$

The OEE calculation factors in the major losses that TPM seeks to eliminate. The first focus of TPM should be on major equipment effectiveness losses, because this is where the largest gains can be realized in the shortest time. The 11 major areas where losses occur are listed in Table 2-2 within four broad categories.

The contemporary business environment of today is characterized by turbulence and aggression, even preemptive practices. In such trying times,

Table 2-2
Eleven Major Losses

Planned-shutdown losses	Performance efficiency losses
1. During breaks, and/or shift changes	7. During minor stops (< 6 minutes)
2. During Planned Maintenance	8. From reduced speed or cycle time
Downtime Losses	**Quality Losses**
3. From equipment failure or breakdowns	9. From scrap product/output
4. From setups and changeovers	10. From defects or reworked items
5. From tooling or part changes	11. During yield or process transition
6. During start-up and adjustment	

organizations are hard pressed to enhance their capability to create value for customers and improve the cost effectiveness of their operations. Maintenance, as a critical support function in businesses with significant investments in physical assets, plays a major role in meeting this tall order. A consensus of a number of surveys indicates that within the manufacturing industry, maintenance spending ranges between 14 and 25% of the total factory operating costs. Within process industries in general, and refineries specifically, the maintenance and operations departments are nearly always the largest with each comprising between 28 and 33% of total staffing. Attaining the right mix of physical assets and making the best use of those already in place to meet business needs are the ways maintenance can contribute to improving competitiveness of physical asset, capital-intensive organizations.

Some of the more significant developments of the last half of the twentieth century that have made the performance level demanded of maintenance ever more challenging are:

- *Evolving trends of operation strategies*: A shift of emphasis from volume to quick response, elimination of waste and defect prevention. With the elimination of buffers in such demanding environments, breakdowns, speed loss and erratic process yields create immediate problems for the timely supply of products and services to customers. Clearly, installing the right equipment and facilities, optimizing the maintenance of these assets and effectively deploying the manpower to perform the maintenance activities are crucial factors to support these emerging trends in operating strategies.
- *Hardening of societal expectations*: A growing, widespread expectation for protection of the environment and safeguarding people's safety and health. A wide range of regulations have been enacted to control industrial pollution and prevent accidents in the workplace. Scrap, defects and inefficient use of materials and energy are sources of pollution, often as a result of operating plants and facilities under less than optimal conditions. In chemical production processes, a common cause of pollution is the waste material produced during the start-up period after production interruptions. Catastrophic failures of operating plant assets and production machinery are also major causes of industrial accidents and health hazards. Maintaining facilities in optimal conditions and preventing failures are effective methods for meeting the evermore demanding societal challenge of pollution control and accident prevention. These are all elements of the core maintenance functions.
- *Technological changes*: Technology has always been a major driver of change and has been evolving at a breathtaking rate in recent decades, with no signs of slowing down in the foreseeable future. Maintenance is

no exception in being under the influence of rapid technological changes. Technologies such as nondestructive testing, transducers/ sensors, vibration measurement, thermography, ferrography and spectroscopy make it possible to perform nonintrusive inspection. The condition of equipment can be monitored continuously or intermittently while operating, through the application of these technologies. The growth of these, and related technologies has given birth to condition, the based maintenance, an alternative to the classical, time-driven approach of preventive maintenance.

Power electronics, programmable logic controllers (PLCs), computer controls, transponders and radiofrequency (RF) telecommunications systems are increasingly being introduced to replace electromechanical systems and hard-wired data links. They offer the benefits of improved reliability, flexibility, compactness, lightweight and often lower cost. Flexible manufacturing cells and computer-integrated manufacturing systems are gaining broad acceptance in the manufacturing industry. In the electric utility industry, automated systems are being installed to identify and deal with faults in the transmission and distribution network remotely.

The deployment of these new technologies has been instrumental in enhancing system availability, improving cost-effectiveness of operations and delivering better and more innovative services to customers. The overall trend continuously presents new challenges to maintenance. New capability has to be developed to commission, operate and maintain systems with new technologies. During the phase-in period, interfacing old and new plant and equipment is another challenge to be handled by maintenance.

- *Changes in organizational systems and the attitudes of people*: In the past, companies were busy producing standard goods and services to satisfy the insatiable demand of their customers. These companies were protected from the onslaught of outside competition through regulation or imposition of trade barriers in their home market. Product life cycle was long due to slower technological change and higher tolerance of accommodating customers who would take what was available on the market. On the human dimension, people perceived work merely as a means to earn a living. All of these conditions have changed in today's tumultuous environment. People at work (the individuals who make things happen in organizations) have undergone significant transformation. The social and demographic changes that have taken place in the current era affect how society regards and defines work. Two examples of these changes are improvements in education and increased faith in the ability of individuals to self-manage their work. Faced with this new reality, progressive organizations are exploring new directions in their labor–management agreements. This in turn has led to the appearance of a number of innovative and often highly successful organizational forms like more horizontally structured organizational patterns, networked organizations, self-managing work teams, virtual organizations and strategic organizational alliances. Within these variations lie many appropriate options for meeting today's challenge of providing excellent maintenance services to internal organization customers.

Figure 2-4 shows the input–output model that models maintenance as a transformation process encapsulated in an enterprise system. The resources deployed to maintenance include labor, materials, spares, tools, information and money. The way maintenance is performed will influence the availability of production assets as well as the volume, quality, cost of production and safety of the operation. Together, these will determine the profitability of the enterprise.

The required investment to implement TPM is very high. However, to offset the very high investment is an even higher return-on-investment (ROI). Through TPM's cooperative effort, job enrichment and pride are created. The new attitudes of TPM dramatically increase productivity and quality, optimize equipment life cycle cost and broaden the base of every employee's knowledge and skills. A word of caution, however; TPM cannot be applied to unreliable equipment. As a result, the company's first investment in TPM must

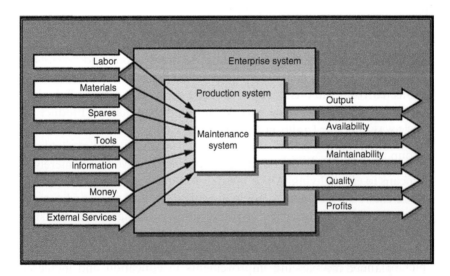

Figure 2-4 Maintenance: Hub of Internal and Enterprise-wide Outputs

include the expense of restoring equipment to its proper, reliable condition and then educating personnel in the proper use and care of their equipment.

In TPM, there are clear assignments of responsibility for the three basic maintenance responses: routine, emergency and backlog relief. TPM is also organized to recognize three distinct (separate but mutually supportive) functions so that each basic function receives the primary attention required. These functions are:

- Work execution
- Planning and scheduling
- Maintenance engineering

2.3.1.1 Work Execution

Effective control of the maintenance work execution function depends upon clear accountability for each type of demand placed upon the organization. The three principal types of demand are *routine* or *preventive, emergency* and *planned work*. The most common structure is composed of three major operating groups dedicated to each of the three principal types of demand. The basic concept of this structure is the establishment of two minimally-sized crews to meet the routine and emergency demands, and a larger third group devoted to planned maintenance.

1. The routine or preventive maintenance group is responsible for the performance of all management approved routine tasks in accordance with detailed schedules and established quality levels. Their work
   a) is specifically defined,
   b) is performed according to a known schedule,
   c) is performed in a planned pattern,
   d) involves a consistent work content and
   e) requires a predictable amount of time.

The group is not interrupted by emergencies or backlog, thereby protecting the integrity of the preventive maintenance schedule.

2. The emergency group has the responsibility of handling essentially all emergency demands, using assistance only when necessary. This allows the planned maintenance group to apply their labor resources to backlog relief.

3. The planned maintenance group is responsible for all work other than emergency and routine. The group is divided into two crews, one

covering work performed primarily in the shops and the other covering work performed in the field.

2.3.1.2 Planning and Scheduling

The responsibilities of the Maintenance Planner are:

- customer liaison for nonemergency work;
- job work-plans and estimates;
- full day's work each day for each person (capacity scheduling);
- scheduling work by priority;
- coordinating availability of labor, parts, materials and equipment in preparation for work execution;
- arranging for delivery of materials to job site;
- ensuring that all jobs (even low priority) are accomplished;
- maintaining data—records, indexes, charts, etc.;
- reporting on performance versus goals.

2.3.1.3 Reliability Engineering

In general terms, the function of Reliability Engineering is the application of engineering methods and skills to the correction of equipment problems that are causing excessive production downtime and maintenance work. Their responsibilities are to:

- ensure maintainability of new installations;
- identify and correct chronic and costly equipment problems;
- provide technical advice to maintenance and proprietors;
- design and monitor an effective and economically justified preventive maintenance program for the TPM program;
- ensure proper operation and care of equipment;
- establish a comprehensive lubrication program;
- perform inspections of adjustments, parts, parts replacements, overhauls, etc., for selected equipment;
- perform and/or oversee vibration and other predictive analyses;
- ensure equipment protection from adverse environmental conditions;
- maintain and analyze equipment data and history records to predict maintenance needs (*includes selected elements of Reliability-Centered Maintenance (RCM)*);
- monitor the effectiveness of maintenance training.

2.3.2 Lean Refinements

Earlier, while describing the Lean manufacturing environment, it was stated that "perhaps above all Lean is about people." Unlike traditional manufacturing organizations, people are not the problem in a Lean enterprise; they are the "problem solvers." This fact, together with the most basic premises of Lean, i.e., the practice of eliminating waste in every area of production including customer relations (sales, delivery, billing, service and product satisfaction), product design, supplier networks, production flow, maintenance, engineering, quality assurance and factory management are the driving forces for the refinements needed to make Total Productive Maintenance Lean.

2.3.2.1 Reliability Excellence

The first refinement involves "raising the bar." Mediocrity is not an acceptable status for the Lean Maintenance Operation. Defining benchmarks for maintenance practices and then meeting (or exceeding or even creating higher benchmarks) like benchmark levels of Best Maintenance Practices (BMP) will be a top priority. What are BMP and how do you go about defining the benchmarks? BMP are established standards for the performance of industrial maintenance. Measuring a plant's existing maintenance operation using the yardstick of BMP can reveal both the degree of maintenance impact on reliability and also permit identification of the specific maintenance processes causing variations in equipment reliability. Following are just a few examples of best maintenance practice standards. (see Appendix C for a comprehensive listing).

- PM hours worked as a percentage of available hours
  PM Hours/Available Hours > 30%
- Emergency Work
  Emergency Labor (hours) / Total Maintenance Labor (hours) < 2%
- PM Compliance
  PM's scheduled by week / PM's completed by week > 95%
- PM Effectiveness
- Overtime Maintenance
  Overtime/Maintenance total time < 5%
- OEE

$$\text{(Equipment Availability)} \times \text{(Performance Efficiency)} \times$$
$$\text{(Rate of Quality)} > 80\%$$

(Can vary by industry)

The defined BMP benchmark listing can include as few or as many practices as it takes to provide an accurate, "across the board" measure of performance of the Maintenance Operation. The list is public domain within the plant and the Maintenance Organization's level of achievement of the defined BMP benchmarks is also publicly displayed.

Achieving the level of equipment reliability necessary to support Lean Manufacturing objectives, requires a Maintenance Organization that is characterized as operating as a minimum at the level of Reliability Excellence in their degree of BMP achievement. The bar graph in Figure 2-5 illustrates, on a percentage scale, the classifications at designated levels of achieving BMP benchmarks.

2.3.2.2 Teams and Teamwork

Lean Thinking is also about teamwork and teams provide the basis for teamwork. Equipment reliability becomes everyone's responsibility in the Lean Environment but the areas having the greatest effect on reliability are maintenance and operations. Regardless of the organization details, where the primary work place is on the shop floor, action teams that include maintenance personnel and operations personnel will be formed. Production line operators support the Maintenance Operation through performance of *autonomous maintenance*. Autonomous maintenance is routine maintenance such as cleaning, inspecting, making minor adjustments and lubricating machinery. Conversely, maintenance members of the production line action team will assist in production equipment setup, provide on-the-job maintenance training to operators and assist in problem detection and correction in their assigned production line.

Both maintenance and operations members of these shop floor teams will also be responsible for continuous improvement in equipment reliability and

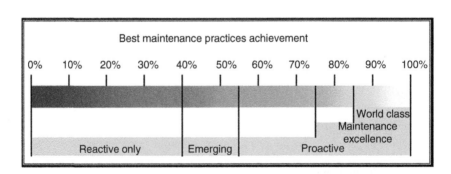

Figure 2-5 Maintenance Excellence and Best Maintenance Practices

product quality. They are responsible for identifying areas of operation that add no value to production/product (waste) and then to eliminate that waste through performance of Kaizen events. Kaizen is the philosophy of continual improvement, putting into practice the belief that every process can and should be continually evaluated and improved in terms of time required, resources used, resultant quality and other aspects relevant to the process. The Kaizen event, often referred to as Kaizen blitz, is a fast turn-around (one week or less) application of Kaizen "improvement" tools to realize quick improvement results.

It is important to note that although "Lean" operating philosophies have been implemented with much success in Japan, their success rate in the United States has been much lower. There are several well-defined reasons for this. The cultures of the two countries are at extreme opposite poles. The mind-set of the Japanese people lends itself to the concepts of Lean practices. Employees in Japan often "volunteer" their own time to the company. Japanese employees often are guaranteed employment "for life." Manufacturing companies in Japan often receive government subsidization, which permits nearly unlimited training, generous employee incentives, few production cost constraints and the ability to "experiment" with a variety of process changes. Company management, even top management, spends more time on the shop floor. When management adopts a vision/mission such as a Lean operating philosophy, the commitment is total, visible and active from the top down.

Conversely, in the Western world in general and the United States in particular, there is a mutual mistrust between labor and management as a result of nearly 200 years of mistreatment of labor by management and of violent revolt by labor against management. Even though these specific situations have not existed for the last 50 years, or longer, the basis for the mistrust is longstanding and ingrained. Witness recently the profit-taking of corporate executives prior to filing for bankruptcy, and the concurrent disenfranchisement of employee retirement plans. The fact that these events have taken place in a statistically negligible number of U.S. corporations has little impact on reducing the indelibility of the attitude reinforcement in the minds of labor.

All too often, companies implementing Lean practices unconsciously place limitations on the degree of implementation, which doom the Lean transformation to failure. There are an enormous number of studies, assessments and audits of unsuccessful Lean implementations that have identified a number of common practices leading to those failures. Among the most common are the following:

- Suggestions for improvement must be submitted to Team Leaders for presentation to management.

- Periodic meetings outside of working hours are held by management to elicit ideas for improvement, but management offers no encouragement for employee input.
- No action is taken on suggestions for improvement.
- There is no visible commitment from upper management for the Lean implementation.
- There are no incentives or rewards offered for successful improvement ideas.
- There is retribution, or publicly made rebukes for unsuccessful or illogical suggestions.
- There is no feeling of involvement or participation on behalf of the workers.
- Empowered team "leaders" are set up by management as "watchdogs."

Clearly, these practices and attitudes will doom any Lean initiative to failure. As stated earlier, Lean is about people. Additionally, Lean is about changing attitudes and it is about total corporate commitment—from the very top management levels to the shop floor—to the concepts of the Lean Operation.

2.3.2.3 New Roles for Managers and Supervisors

With the formation of action teams, which are empowered, self-directing and team activity oriented, the roles of management and supervision require rather dramatic changes to take place. Instead of *directing and controlling*, the new role is one of *support*. In their new roles, managers and supervisors provide overall guidance for the work that is clear and engaging. They also offer hands-on coaching and consultation to help employees avoid unnecessary losses of effort, to increase task-relevant knowledge and skills, and to formulate uniquely appropriate performance strategies that generate real process improvements.

2.3.2.4 Organizational Focus

One of the dominant characteristics of Lean Organizations as compared to traditional organizations is the flattening of the organization's structural hierarchy with fewer layers of middle-level management. This is a direct result of creating and empowering the self-directed action teams. Along with the flatter structure, a shift in focus needs to occur within the organization if empowerment and continuous improvement are to be sustained. The emphasis now is on recognition of the employee as the plant's most valuable asset. Among the refined characteristics of the Lean Organization are:

- Top management is commited to teamwork and the concept of team-based rewards and recognition.
- Management is available and visible.
- The organization relies on structured processes, policies and documentation.
- A performance measurement system is in place.
- A strong network is in place for vertical, horizontal, diagonal, intrateam and interteam communication.
- Employees are regarded as the organization's most valuable assets.
- Employees value empowerment and involvement as a form of reward and recognition.
- Employees participate in training.

2.3.2.5 Expanded Education and Training

Educational resources, which can include technical consultation as well as training, are available and accessible to employees with identified needs. For instance, the specialists of maintenance department are called upon to upgrade operators to autonomous operator–maintainers. In the Lean Environment, the training is not being limited to transfer of technical skills and knowledge needed for optimal task performance. It also covers generic matters like the business imperatives peculiar to the organization (what determines the value of its product and services to customers), problem-solving techniques, team dynamics and facilitation skills. Additionally, training for managers addresses issues such as their new roles (leader, communicator, coach, resource provider) and the new management behavior that will align efforts and generate commitment for organizational goals.

2.3.2.6 Maintenance Optimization

The fundamental premise of Lean Operations is the elimination of waste. One of the larger categories of waste in the Maintenance Operation is the performance of unnecessary maintenance. In the Lean Enterprise, Maintenance Engineering and the Maintenance Planner have defined roles for optimizing maintenance. Figure 2-6 graphically illustrates the optimized maintenance interval (frequency) and the effect on maintenance costs that the wrong (nonoptimized) maintenance interval has. In addition to optimizing maintenance intervals, there are a number of considerations in Maintenance Optimization. They include the use and/or application of:

- Predictive Maintenance (PdM);
- Condition Monitoring (CM)
  (this includes Equipment Testing, Component Inspection, Equipment, Component and Process Measurement (speed, size, flow, operating range, etc.) utilizing installed instrumentation (Calibration Checks, etc.);
- analysis, evaluation and adjustment of corrective maintenance and/or overhaul criteria;
- addition and/or Deletion of equipment in PdM or CM program (failure analysis);
- addition and/or deletion of PdM technologies.

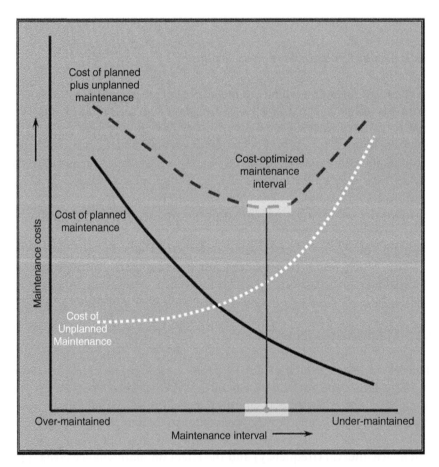

Figure 2-6 Optimized Maintenance Interval and Maintenance Costs

Other refinements to Maintenance Operations wrought by Lean Thinking involve other elements of the manufacturing organization and thereby affect the ways in which the maintenance branch interfaces with those elements. As the text delves deeper into maintenance planning and scheduling, these refinements together with the ways in which they affect the planning and scheduling functions will be described in detail.

2.3.3 Lean Maintenance: Prerequisite of the Lean Plant/Facility

In the Lean Manufacturing Plant, production equipment reliability is critical to achieving the "Lean" state or condition. It is for this reason that creation of a Lean Maintenance Operation must be completed before the remainder of the functional elements within the plant (in particular operations) can undertake Lean Operations. The Maintenance Organization has gained new recognition in the lean environment. While the importance of maintenance has always been high, it was not always given the kind of attention that was deserved unless there was a failure related production stoppage. The advent of Lean Thinking has raised the bar for the maintenance operation and its practitioners. Responding to the new level of attention is best done with professionalism, enthusiasm and continual learning through classroom, doing, observing and recognizing new and better ways of performing. Lean and the Lean enterprise are not passing fads; lean is here to stay and it is a must for the manufacturer who wants to compete in the twenty-first century.

3

Governing Principles and Concepts of Lean Maintenance

It was stated in Chapter 1 that the most basic or underlying premise of Lean Operations is the elimination of waste. As a basic premise, or basis of operations, it is also a governing principle, or ideal of Lean Operations. Governing principles are the applied standards for defining policy, direction and objectives for the lean enterprise. There are three governing principles behind Lean Manufacturing:

Waste Elimination—Actively seek to identify and eliminate waste, which is anything (any part, practice, process, design element, work environment, organization element or policy) that does not add value to or for the customer.

Focus on Customer—The operation is focused on what the customer values, which in general are related to low cost, product quality and reliability, timeliness and demand for a product/function.

Quality Generated at the Source—Quality is built in, not inspected and/or tested in. That is, the production equipment and manufacturing processes generate first quality components and products the first time through; rejects, rework and work-arounds do not exist.

If these are the governing principles of a Lean Manufacturing plant, what are the governing principles of the Maintenance Organization in the Lean Manufacturing Plant? Since the function of maintenance is to support operations, do the same underlying standards apply?

3.1 LEAN MAINTENANCE GOVERNING PRINCIPLES AND CONCEPTS

The governing principles of every organizational element must be the same as those of the parent organization or corporation. In diversified enterprises that include, for example, manufacturing, service operations (e.g., financial services), facility operations (e.g., hospitals) and similarly diverse business units, individual organizational elements will have as their governing principles a subset of the parent organization's governing principles. Divergent principles create divergent policies and organizational elements risk opposing the goals and objectives of other interrelated, even interdependent elements. Thus, the governing principles of the maintenance operation, although focused somewhat differently, remain the same as those of the plant itself. To reflect the different focus, the governing principles of maintenance can be restated as:

Waste Elimination—Actively seek to identify and eliminate waste, which is anything (any part, practice, maintenance process, tools or instruments, "bench" spare, waiting time, travel time (e.g., walking from one job to the next at opposite ends of the plant), work environment, organization element or policy) that does not add value to or for the customer. The maintenance worker's customer is most often operations but also still includes the buying customer of the plant output.

Focus on Customer—The maintenance operation is focused on what his or her customer (operations) values, which is primarily equipment reliability. The operations customer also values timely completion of scheduled maintenance, restoration of equipment to "ready-to-produce" condition at completion of maintenance and similar production specific needs from maintenance activities. This principle also provides value to the end customer buying the products output by the plant through support of the production process.

Quality Generated at the Source—The quality of maintenance activities translates to correctly performing the right maintenance at the right time and the first time (no repeats of maintenance tasks because the tasks are performed correctly the first time). Maintenance tasks are performed using the correct parts, materials and consumables in the correct quantity. Maintenance activities yield production equipment that manufactures first quality components and products the first time through; rejects, rework and work-arounds do not exist.

As these restated governing principles show, the emphasis is the same but the focus is placed on the immediate maintenance customer—operations.

Successful maintenance functions are built upon a foundation of governing principles and a layer of support concepts—concepts of how the Lean Maintenance Operation functions. Ten concept statements follow, which are considered essential for support of the Lean Maintenance Operation. Individual Maintenance Operations may have additional concepts of how maintenance is performed in their plant, however they must still support the corporate mission and vision as well as the maintenance mission.

1. Management has developed and communicated a clear statement of the maintenance mission to ensure that departmental activities are consistent with and supportive of the facilities' operating plan and associated operating strategies.
2. The concept of operating custodianship is basic to facility philosophy.
3. Responsibility for determining how best to satisfy maintenance needs rests with the maintenance department through technically qualified knowledge and advice.
4. Maintenance of processes, equipment and facilities is a responsibility shared by all units of the organization.
5. Maintenance is a cornerstone of the operation. Policy statements are issued by the management to ensure that the maintenance program is an integral part of the operating strategy. Thus, common understanding of the maintenance effort is assured and its procedures are rooted to these facility-wide practices.
6. Maintenance philosophies and functions supporting facility strategies are written within a procedures manual and effectively communicated to all concerned departments.
7. Effective maintenance is proactive, not reactive. When failures occur, preventive measures are sought to preclude recurrence.
8. There is a clear definition and understanding of responsibilities (primary and supportive) between the operating (production) and maintenance departments.
9. Mutual understanding and cooperation is excellent. Important procedures, such as the work order system are followed with reasonable uniformity.
10. The proactive nature of the Total Productive Maintenance (TPM) program is not overlooked when the plant is under pressure to meet output targets.

3.1.1 Vision and Mission

It is important to understand the relationships between governing principles, objectives and goals as well as an operation's vision and its mission (Figure 3-1). The underlying drive for the way you do things, whether in life or in business, are your particular set of principles and concepts. Everything else grows from these roots. A vision is a portrait of what you would like ultimately to become, or in business, where the organization would like to be in the future. Your mission is the purpose of your existence or of the Maintenance Operation's existence. Mission is the reason you do what you do. Goals are the measurable steps toward fulfilling the mission while objectives are the conditions that must be met in order to fulfill the mission. Chapter 1 listed eight objectives of the Lean Maintenance Operation. These are the minimum required objectives. Depending on the Mission Statement of the Maintenance Department, additional objectives may be required.

A Maintenance Mission Statement will clearly state the reasons, or purpose, for the maintenance department's existence as it supports the company's vision and its strategic and operations (production) plan. Vision statements are normally generated only at the top level of the company. They become, by force of policy, the vision of every element within the organization. Maintenance objectives and goals are synchronized with the maintenance department mission statement and will be consistent with the company's strategic plan and operations/production plans, which in turn, are formulated to realize the company's vision. The maintenance mission statement must support the company's stated mission and vision. The following is one example of a maintenance mission statement:

"Our maintenance mission is the provision of timely, high quality, cost-effective service and technical guidance to support short-range and

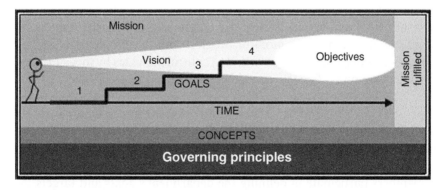

Figure 3-1 Relationships: Mission–Vision–Objectives–Goals

long-range operating/production plans. We will ensure, through proactive practices rather than reactive, that assets are maintained to required levels of reliability, availability, output capacity, quality and customer service. The maintenance mission seeks to continuously improve work practices and maintenance effectiveness and will actively seek out and eliminate waste. Our mission is to be fulfilled within a working environment that fosters the highest levels of safety, morale and job fulfillment for all members of the maintenance team while protecting the surrounding environment."

3.1.2 Strategic Plans, Goals and Targets

Strategic planning is the weighing of and choosing among long-term alternatives to develop a timeline of action, milestones and goals necessary for a company to achieve elements of its vision. Strategic plans are more focused on marketing, sales, product development, production and profitability aspects of the company vision statement. Strategic plans emanate from the top level of corporations. They often require subsidiary elements of the company to develop a strategic plan of their own detailing their action requirements to support the corporate level strategic plan. At a minimum, a set of timelined goals and targets necessary to support and achieve the short- and long-term milestones and/or goals of the corporate level or immediate client support strategic plan are required. At the level of the Maintenance Department, the latter requirement of goals and targets in support of the Operations Department's strategic plan is normally applicable.

In the Lean Environment, the Maintenance Department will define specific goals for achievement in relation to the Operations Department's strategic plan. Additionally, targets are set for maintenance performance in terms of equipment up time, maintenance costs, overtime, work-force productivity, supervisor's time at job-sites and similar measures. Such specific targets as these enable management to monitor progress and the effectiveness of the maintenance management program and to control activities by focusing corrective attention on performances or levels that consistently fall short of the targets.

Superficial goals lead to superficial results. A clear understanding of the mission is critical to success. Similarly, unachievable goals will frustrate maintenance supervisors and workers and ultimately lead them to abandon the pursuit of goals and targets. In the Lean Environment, maintenance staff members participate in defining the department's goals and targets. A positive and productive mind-set results in positive and productive performance. The Lean Environment builds a climate that promotes self-motivation.

Most targets are merely mile-markers for goal attainment. For example, existing overtime hours are 11% of straight-time hours. The goal is to reduce overtime to 5% or less within 6 months. Targets are established at 1-month intervals of 10%, 9%, 8%, etc. Once defined, performance measurement strategies are defined, the goals and targets are published and measured performance is tracked against target values in a visual, public display (Figure 3-2).

Nearly as important as goals and time-targeted achievement levels are the metrics, or measurement processes, necessary to determine progress towards performance goals. Without the measure of progress, course adjustments cannot be made in a timely fashion to correct deviations from targeted paths. Rewards for meeting target performance levels are just as difficult to meter out without the benefit of performance metrics. The Information Technology (IT) department has an important role in identification of data sources and creation of performance indicator reporting formats.

The primary source of maintenance performance data is the Computerized Maintenance Management System (CMMS). The accounting system is also an important source of cost and labor data. In those companies utilizing a fully integrated Enterprise Asset Management (EAM) system, the acquisition of data as well as creation of performance indicators and reports is greatly simplified. An EAM system is a plant-wide (or multisite wide) information management system that encompasses data management for all departments in the organization. The most common maintenance performance indicators can be classified into three categories:

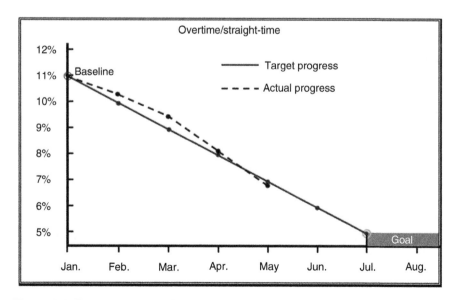

Figure 3-2 Typical Display of Goal and Performance

1. measures of *equipment performance*, such as availability, reliability and overall equipment effectiveness;
2. measures of *cost performance* such as labor and material costs of maintenance;
3. measures of *process performance* such as schedule compliance and the ratio of planned and unplanned work.

3.2 OPERATIONS AND MAINTENANCE

In the pre-Lean world, there was not much integration between maintenance and operations functions, and most of what existed was between maintenance planning and scheduling, and production scheduling. The other primary interfacing was initiated through trouble and failure reports, and work requests initiated predominantly by production line operators.

In the Lean Environment, the integration of maintenance and operations is much tighter. The tighter integration begins with the formation of empowered action teams consisting of both operators and maintenance technicians and sometimes includes a maintenance engineering representative. Depending on plant size, layout and other factors, teams are organized by plant area, production line, process or stage, or facility zones. Each team is accorded stewardship of the plant assets within their area assignment. (*Stewardship: the conducting, supervising or managing of something; especially: the careful and responsible management of something entrusted to one's care.*)[2]

3.2.1 Operations, Maintenance—A Partnership with Two-Way Responsibilities

Along with the formation of the Lean action teams comes a new (although it should have always existed) set of responsibilities between team members. The strict separation of production work and maintenance work no longer exists, although the new functions and responsibilities must be well defined. In the Lean Environment, new functions include:

- operators performing equipment cleaning;
- operators being a part of "early warning" system–in order to perform daily equipment checks;

---

[2] *Merrian-Webster's 11th Collegiate Dictionary.*

- team members pitching in to restore equipment when they spot any problem;
- operators performing routine (autonomous) maintenance;
- training and certification of operators performing autonomous maintenance to perform designated routine maintenance tasks;
- maintenance personnel assisting in set-up following maintenance work;
- considering action team members as qualified personnel, and not as castoffs;
- operations and maintenance supervisors spending more time at the job site;
- performance of major maintenance work requiring highly skilled crafts, by centralized (non-team) maintenance forces and
- effective communication between team members: communicate, communicate, communicate!

In addition to these new functions and relationships between members of Lean action team, there is a new (although always existing) set of responsibilities, both shared and to each other, for operations and maintenance departments in the Lean Environment. These responsibilities should be formalized through inclusion in departmental standard operating procedures (SOPs).

Shared Responsibilities:

- Both must work cooperatively to reduce emergencies—converting costly breakdown to proactive maintenance.
- Both must actively pursue the identification of waste and then jointly seek ways to eliminate it.

Responsibilities of Operations to Maintenance:

- Operate machinery and equipment properly.
- Know the conditions and performance of equipment assets. Maintain surveillance and establish awareness of system and equipment in order to detect unsatisfactory or anomalous conditions and anticipate essential work. Report such conditions to maintenance for diagnosis and action.
- Request and authorize repairs, replacements and alterations and describe them clearly in writing.
- Avoid unnecessary work and fictitious priority. Help to control the volume variance within maintenance budgets.
- Accept equipment ownership (stewardship).
- Participate in performance of equipment maintenance work to the degree specified, authorized and trained.

- Plan for and provide adequate equipment access for timely performance of programmed and scheduled maintenance.
- Communicate capacity needs.

Responsibilities of Maintenance to Operations:

- Based on authorized orders for maintenance or repair service, define and execute the required work in a timely fashion, with quality workmanship, knowing what is to be done, when and how best to do it. Then do it right.
- Assist operations in establishing a practical level of maintenance so that long- and short-term production plans can be met and repairs can be anticipated, planned and scheduled.
- Maintain facilities at specified levels of operating condition, at lowest possible cost consistent with the goals of producing quality products as economically as possible.
- Actively participate with operations to execute and continuously improve a comprehensive preventive maintenance program.
- Convert emergency work to planned work by anticipating it.
- Make repairs and replacements at intervals required for optimal operating efficiency and in a manner creating as little production loss as possible.
- Constantly strive to improve maintenance work methods, completeness and neatness with the goal of quality work at minimal cost.
- Effectively plan, schedule and coordinate maintenance work with operations in advance, to permit them to plan for out-of-service equipment and to minimize nonproductive time and production stoppage.
- Prior to execution, thoroughly review all shutdown work with key operations personnel so their first-hand operating knowledge can be fully utilized.
- Provide regular feedback regarding status of work requests and completion promises.
- Advise operations as to the levels of risk and the potential costs related to operating equipment believed to be close to failure. Develop techniques for predicting failure of critical facilities with reasonable accuracy.
- Inform operations of systems or equipment requiring excessive maintenance and take appropriate action to reduce it.
- Account for the level of cost incurred in the performance of requested maintenance (standard vs. actual—the performance variance).
- Regard operations as a customer (internal).

3.2.2 A Lean Mandated Marriage

You will recall that the underlying or governing principles of Lean Thinking are:

Elimination of Waste, Focus on the Customer and Quality the First Time (at the source). These principles dictate the tighter integration of operations and maintenance. Look at the six focal points of Lean action team activities practicing TPM.

1. activities to optimize overall equipment effectiveness and reliability through stewardship;
2. elimination of breakdowns through a thorough system of maintenance throughout the equipment's entire life span;
3. autonomous operator maintenance (this does *not* imply that operators perform *all* maintenance),
    (a) use lower-skilled personnel to perform routine jobs that do not require skilled craftsmen;
    (b) use operators to perform specific routine maintenance tasks on their equipment;
    (c) use operators to assist mechanics in the repair of equipment when it is down;
    (d) use computerized technology to enable operators to calibrate selected instrumentation;
    (e) use maintenance technicians to assist operators during shutdown and start-up;
4. day-to-day maintenance activities involving the total work force (engineering, operations, custodians, maintenance and management);
5. company-directed and motivated, yet autonomous small group activities. Small group goals to integrate with company goals;
6. continuous training,
    (a) formal (classroom/lab with mock-ups);
    (b) On Job Training (OJT);
    (c) one-point lessons;
    (d) team members train each other.

It is clear that these activities are primarily focused on elimination of waste, e.g.,wasted time, waste from unreliable equipment, waste through over-qualified labor performing routine, low-skilled activities, etc. They also focus on quality at the source through maintenance of equipment tolerances and precision, and finally they are customer focused. Maintenance is focused on supporting operations (their direct customer). Operations is focused on their customers: operations—by enabling execution of short and long-term operating plans; consumers—through lowest cost and highest quality production runs. Thus, the cooperative, coordinated and focused joint efforts of maintenance and operations are essential for success in the Lean Environment.

The objective of Lean Maintenance is not to minimize the maintenance budget, but to increase profitability through efficiency, reliability and customer satisfaction. It is not necessarily to minimize input or to maximize output, but to optimize their combined impact on profit. More than sheer quantity, "output" includes improving quality, reducing costs and meeting delivery dates while increasing morale and improving safety and health conditions as well as the working environment in general.

To achieve a truly Lean Maintenance Operation, the organization must embrace the concept of Reliability Excellence (Rx). Rx is a process that ensures the entire organization—Engineering, Production, Procurement, Maintenance, and others—is focused on achieving and preserving equipment reliability.

Experience has shown that in order to achieve excellence in any endeavor, there are five phases that must be addressed in turn. These are illustrated as the levels to excellence in Figure 3-3, The Excellence Model.

The "Principles" level provides necessary support to all subsequent levels. This is the point at which the organization's senior management determines the philosophies that will govern site behaviors. It contains both a vision and values component. Creating an environment of performance that has a clear and consistent vision and allows the employees to maximize job satisfaction provides the foundation upon which excellence is built.

The second level, "Culture," is where the working environment is established. Where the "Principles" defined the vision and mission, the culture

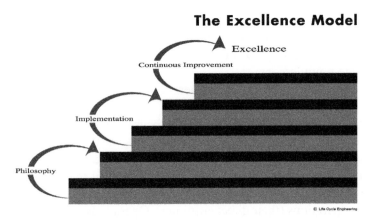

The Excellence Model

Excellence

Continuous Improvement

Implementation

Philosophy

© Life Cycle Engineering

Figure 3-3 The Excellence Model

defines the system in which the work shall be done. To a great degree, it is where the principles are put into actual practice.

The third level is where the "Processes" are put into place. Processes are the methods by which work gets done every day. These will inherently reflect the culture that has been established below; if a disciplined culture is in place, processes will likely be followed with uniformity.

The fourth level, "Optimization," builds on the work processes themselves by addressing efficiency and effectiveness issues. It is where items such as technology and control mechanisms are applied to enable the processes to reach their maximum potential.

Finally, the top level is where "Sustainability" is created. It is the continuous improvement loop that reinforces all previous levels and ensures that incremental performance gains are not lost. Excellence is not a static standard. The bar is always being raised; so sustained peak performance requires continuous improvement.

Excellence does not occur by chance, nor does it depend upon superhuman efforts by selected individuals. It is the collective result of getting the details right, quantifying those actions, and doing them better each and every time. The excellence model, while presented in layers, is not meant to imply that these actions can only occur in sequence; however, it is important that any transformations start at the foundation level. This ensures that management commitment is available to prevent the organization's internal defenses from completely derailing the efforts.

Lean Maintenance organizations achieve Reliability Excellence by applying the Excellence Model concepts to everything affecting the facility's physical assets. Management philosophy drives the necessary culture to support high levels of equipment reliability. Work processes are in place that will ensure high levels of reliability. These processes are optimized to achieve maximum efficiency and effectiveness. Programs are in place to sustain gains in reliability once they are achieved. Specific elements that are required to be in place for each level are illustrated in Figure 3-4, The Reliability Excellence Model.

The "Principles" level consists of two elements, Plant Partnerships and Management Commitment. The philosophy that must be in place to support Reliability Excellence demands that *all* functions in an organization are responsible for reliability, just like safety. Each department must recognize the impacts their actions have on reliability. Management must foster an environment where each department functions as equal partners in achieving the organizations mission. Management must also create and maintain an effective culture of discipline to achieve high levels of reliability.

The culture is the set of rules and philosophies that govern behavior on a day-to-day basis, and the following elements are necessary to support Reliability Excellence:

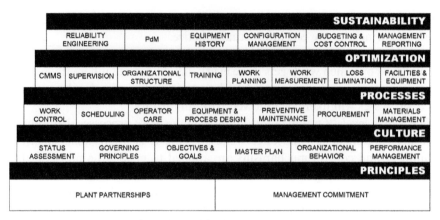

Figure 3-4 The Reliability Excellence Model

- A periodic *Status Assessment* to develop a clear and objective view of the current state of organizational performance. This is required to establish a baseline for continuous improvement.
- A written set of *Governing Principles and Concepts* that represent a set of "shared beliefs" that all functions in the organization must have to support Reliability Excellence. In effect, these become the rules and attributes defining the new culture.
- A set of *Objectives, Goals and Targets* that are derived from the organization's mission. These are the critical measures that drive the appropriate behaviors required to achieve Reliability Excellence.
- A living *Master Plan* that defines the activities and initiatives required to implement Reliability Excellence. Typical organizations require three to five years to make the transition from a reactive culture to a proactive one, and a written project plan is essential to help the organization sustain its focus for this length of time.
- A working environment that promotes the correct *Organizational Behavior*. Proactive organizations are highly disciplined and exhibit a concern for precision in everything they do.
- A *Performance Management* system that enables management to hold individual employees accountable for results. Each employee must have a "balanced scorecard" that provides objective measurement of their individual contribution and allows them to be rewarded for innovation and creativity.

When a supportive culture is established, the processes that produce Reliability Excellence must be put into place. These include the following:

- An effective *Work Control Process* that is the vehicle by which all work is managed and documented. It ensures that any physical work done to the assets is value-added, performed efficiently, costs are effectively controlled, and useful history is recorded.
- A weekly *Scheduling* process that allows coordination of resource availability and equipment access. Weekly schedules are developed jointly between the users and maintainers of the assets to ensure all labor, materials and contractors are available when the equipment is available, and any delays in work execution are avoided.
- An *Operator Care* process that holds the users of the assets accountable for basic, routine equipment care. Reliability Excellence requires that operators accept ownership for the equipment performance and are the first line of defense against failure.
- An effective *Equipment and Process Design* system that focuses on minimizing the total life cycle cost and maximizing asset reliability. No amount of maintenance effort can inject reliability into an asset that is of marginal design.
- A *Preventive Maintenance* process whose objective is to prevent equipment failure and the corresponding downtime and repair cost. This process ensures that basic cleaning, inspection, and lubrication required for equipment health are done at the appropriate time.
- An effective *Procurement* process that ensures materials are purchased with an eye toward minimizing the total life cycle cost of the asset. Significant hidden costs (maintenance, training, spare parts) can be experienced if assets are procured on the basis of low initial price alone.
- An effective *Materials Management* process that ensures the right materials are available at the right time in the right place at the right price. To effectively fulfill its mission, the maintenance function is dependent upon reliable and prompt material support (spares, replacement parts, supplies and special tools).

When the basic processes are in place and functional, optimization activities can occur. This level is typically where technology is applied to enhance efficiency and effectiveness. This level consists of the following elements:

- A *Computerized Maintenance Management System* (CMMS) to automate the work process and enable collection, dissemination, and analysis of data. Lean Maintenance organizations run on information, and the CMMS is the primary source of this information.
- Good *Supervision* of productive work execution. Effective supervision is the art of getting average people to produce superior work. The supervisor is the direct linkage between the hourly workforce and

management; it is his or her responsibility to enforce the appropriate organizational behavior.

- An effective *Organizational Structure* that promotes a proactive mindset. The structure should allow the three types of Lean Maintenance work (scheduled preventive and predictive tasks, planned backlog jobs, and response to legitimate emergencies) to be accomplished in an efficient fashion. It must also contain provisions for the three forms of effective management: planned work preparation, effective workforce supervision, and engineering dedicated to the elimination of repetitive failure.

- An effective *Training and Workforce Development* program that continually enhances craft skills. Skill development is required to keep pace with advancing technology and to reinforce the precision mindset needed for a Lean Maintenance culture. The program should be targeted at individual development needs defined by well-designed skills assessments.

- A *Job Planning* function that maximizes the productivity of the craft workforce. An effective planning function eliminates unnecessary waste from the work process by ensuring that all materials, tools, support services, and technical information are assembled before the job begins, enabling crafts to perform the work with minimal delay.

- A *Work Measurement* system that enables reasonably accurate job estimates to be produced with minimal effort. A Lean Maintenance organization must be able to make accurate promises when taking machines out of production for maintenance or repair. Effective estimating also allows the Lean Maintenance organization to staff work crews with the right number of craft resources.

- A *Loss Elimination* program that allows the organization to measure its capacity losses and set actions in place to reduce or eliminate them. Lean organizations strive to balance capacity with demand. Understanding the causes of loss in potential capacity enables the organization to eliminate them and to take advantage of additional market share when the opportunity arises.

- The appropriate *Facilities and Equipment* that allow and encourage precision workmanship. A clean, well-lit work environment is crucial to enabling quality equipment rebuilds, and craft technicians must have the necessary precision tools to foster the proactive mindset needed for Reliability Excellence. Adequate space is needed for training, meetings, support staff offices, and storage of precision tools and critical records.

The final level implements those elements necessary to sustain Reliability Excellence and enable continuous performance improvement. There will always be opportunities for reliability improvement, and the systems and structure necessary to manage these must be in place and functioning smoothly. This level consists of the following elements:

- A sound *Reliability Engineering* function that is responsible for driving out sources of repetitive failure. Its mission is to provide the proactive leadership, direction, single point accountability, and technical expertise required to achieve and sustain optimum reliability, maintainability, useful life, and life cycle cost for a facility's assets.
- A *Predictive Maintenance* program that enables detection of problems early enough in the degradation process to allow advanced preparation for corrective action. It should also include the capability for analysis of the operating dynamics of machinery to identify issues related to design, maintenance, or operational procedures that impact equipment reliability.
- The use of *Equipment History* to identify repetitive problems and to enable optimization of the Preventive and Predictive Maintenance programs. It also enables refinement of the Planning and Work Measurement processes by providing actual information that can be compared to initial estimates.
- An effective *Configuration Management* process that provides for engineering review of proposed equipment modifications to identify any potential unintended consequences. It also ensures that necessary information—O&M manuals, Bills of Material, repair procedures, etc.—remain current when there is a change to the equipment. In this fashion, it provides sustainability to the critical documentation developed in the elements previously discussed.
- A *Budgeting and Cost Control* process that enables effective management of equipment maintenance expenditures. Since the Maintenance and Repair budget represents one of the single largest controllable cost line items in any plant's budget, it justifies more effort to develop than simply factoring last year's results by some arbitrary percentage. An effective budgeting process considers individual equipment maintenance needs, and therefore should be developed from a "zero" basis by major equipment item.
- A *Management Reporting* process whereby management receives feedback on performance of a given process and makes corrections as needed. An effective reporting system sustains the Reliability Excellence initiative by tracking performance against those critical metrics defined in the Objectives, Goals and Targets. It also supports continuous improvement by validating that desired changes in organizational or process performance are actually occurring.

The five levels of Reliability Excellence must be addressed in turn for a site to fully achieve its potential in reliability, low cost, and profitability. Each builds upon the successes of the previous level. If one element in a level is not healthy, it jeopardizes the stability of the levels that follow.

3.3 WHY PLAN?

Planning and scheduling maintenance work is not a simple undertaking. Even when the planning is supported by a CMMS or EAM information management computer program, planning is a complex, labor-intensive exercise. So, why bother?

Planned maintenance reduces the wait and delay times that mechanics encounter when doing unplanned work. A major result of failure to plan is that putting out today's fires is given priority over planning for tomorrow, thus ensuring an ample supply of kindling to be consumed in future reactive fires.

Poor utilization of a mechanic's time is usually not his or her fault. Nothing is more detrimental to maintenance performance and morale than poorly planned jobs. All stimulation and enthusiasm is lost when a crew arrives at a job site only to find that the work can neither be started nor completed expeditiously due to lack of managerial foresight and planning. Mismanagement is highly visible.

Conversely, when it is obvious to the crew that management is effectively doing its part towards more efficient operation through well-planned maintenance work, the crew takes pride in its work and organizational performance improves.

Urgency consumes the manager; yet the most urgent task is not always the most important. The tyranny of urgency lies in distortion of priorities— subtle cloaking of minor projects with major status, under the guise of "crisis." The measure of an effective manager is the ability to distinguish important from urgent priorities, refusal to be tyrannized by the urgent and rejection of crisis management methods.

Under all circumstances, when maintenance is performed, it is planned. It is a question of who is doing the planning, when they are doing it, to what degree of detail is it planned and how well, or effective, is it planned. All too often, the wrong person does what little planning there is, at the wrong time and usually under the pressure of a reactive maintenance environment.

3.3.1 Advantages of Maintenance Planning

Without effective planning and scheduling, proper management of the widely varying scope and diverse activities performed by the Maintenance organization is impossible.

When was the last time you saw a company begin production run for a new product without someone planning for

- the material it would take,
- the equipment and set-up required,
- the labor required and
- the time required to produce a finished product.

The very thought of such an unplanned approach is outside the realm of believability. Why should complex, constantly changing, maintenance activities be any different? Table 3-1 also makes an excellent case for the application of Maintenance Planning. These statistics illustrate the potential of a successfully implemented planning function in a plant without increasing total headcount. The result is the potential of a remarkable 77% improvement in labor efficiency!

Additional advantages of planning maintenance activities include:

- work and workload measurement;
- accurate promises;
- better methods and procedures;
- establishment of priorities;
- monitoring of job status;
- coordination of labor, materials, equipment and schedules;
- coordinated trades;
- parts and materials provided;
- production equipment available;
- work schedules (vacation, etc.) and production schedules accommodated;
- bottlenecks and interruption anticipated and avoided;
- preventive maintenance becomes vital part of plan;
- off-site preparation;
- aid for the supervisor;
- maintenance function stature.

Table 3-1
Effects of Planning

	Maintenance Crews A and B	
	Without Planning	**With Planning**
Supervisors	2	2
Tradesperson	20	19
Planner	0	1
Direct Work	35%	65%
Equivalent Labor	7	12.4

There are also advantages for a host of other plant functions:

- Operations benefits with effective planning and scheduling:
  - reduce cost of maintenance while improving service;
  - provide data, expert analysis and advice on maintenance performance;
  - minimize downtime and interruptions to operations;
  - permit operations to describe, approve and control timing of work to be done;
  - render better service to operations by performing most important jobs (as determined by operations);
  - apply technical and maintenance experience to analysis of each job.
  - provide orderly procedures for processing work to prevent orders from getting lost:
  - maintains accurate backlog status;
  - reports completion promptly;
  - provide expert maintenance advice to operations through maintenance planner;
  - provide a single contract for all in process, scheduled and emergency work (area supervisor or functional supervisor);
  - require that operating personnel anticipate repair work before jobs become emergencies;
- maintenance supervisors benefit with effective planning and scheduling:
  - permit advanced determination of labor and time required to complete each job that aids maintenance supervisor in measuring the performance of their crew;
  - allow elimination of delays due to waiting for information, materials, equipment, other skills, tools, etc.;
  - provide an overall plan for the supervisor on which they can base their pre-thinking and pre-planning for day-to-day work and future work;
  - provide the cooperation necessary in connection with each job to do work as prescribed by operations;
  - provide a central source of information concerning maintenance work, equipment and equipment repair;
  - allow the supervisor to devote closer attention to supervision of work in the field or shop;
  - provide special tool and equipment requirements;
  - reduce clerical work for the supervisor;
  - permit advance determination of the number of personnel needed in a given area or location;
  - reduce the number of job interruptions once work is started;
  - establish job goals for the work force;

- predetermine and arrange for required shop work;
- apply specialized ability to the planning and scheduling of shutdowns.

There are also benefits to management with effective planning and scheduling:

- reduce total cost of maintenance while improving conditions of equipment and operating facilities;
- permit accurate forecasting of labor and material needs;
- permit immediate recognition of labor shortages and excesses;
- enable management to level out peak workloads;
- provide factual data required for evaluation of performance and corrective action as needed;
- provide close and constant liaison between operations and maintenance through the maintenance coordinator;
- provide close to 8 hours of productive work for each person and thus increase productivity through the elimination of delays;
- permit more accurate collection and analysis of cost and assure an economic level of maintenance with less interruption of production;
- highlight for analysis, requests for work of questionable justification;
- provide for maximum delegation of authority to permit decisions to be made by those who have firsthand knowledge of the problem;
- provide management with measures of efficiency, which point out variation and outstanding performance.

3.3.2 Objectives and Goals of Maintenance Planning

Of the many activities associated with Maintenance Operations, planning and scheduling have the most profound effect on timely and effective accomplishment of maintenance work. Combined with preventive/predictive maintenance, they provide significant bottom line results. Without them, proper management of the widely varying scope and diversity of activities performed by the maintenance organization is impossible. Maintenance Planning, and coincident scheduling, form the communication center from which all maintenance activity is coordinated.

Delay avoidance is the basic goal of planning and scheduling.

The objective of job planning is to allow maintenance tradesperson to prepare for, perform and complete each job without encountering time-wasting delays, and to see that the job is safely performed to the satisfaction of the requester at optimal cost. Each hour of effective planning typically returns

3 hours in technician time saved or the equivalent savings measured in materials and production downtime. Do not neglect customer relations.

Maintenance Planning, or work preparation, has six primary objectives. These are:

1. optimal support of the operational production plan by improving maintenance in the broadest sense, considering both the technical aspects and service provided to the internal customer;
2. completion of maintenance work when it is needed, in a safe and efficient manner, at the most effective (optimal) cost;
3. minimization of lost production time due to maintenance;
4. optimized utilization of maintenance labor and materials through effectively planned and balanced schedules;
5. equitable resource allocation based on understood criteria and the varying business needs of the internal customers supported;
6. minimization of labor delay and idle time through effective coordination of all participating functions.

4

Origins of the Maintenance Planner

Depending on how you define "Planning Maintenance" and "Scheduling Maintenance," the origin of Planner/Schedulers can vary significantly. Earlier, it was pointed out that pre- and post-flight checks on military, and commercial aircraft had their origins in First World War. The aircraft branches of the military continued to lead the evolving practice of maintenance with their adoption, in the 1930s, of standardized, periodic inspections—the 30-, 60-, 90-, 120-hour (flight time) inspections of airframes and aircraft engines. Each increasing hourly-based inspection involved an increasingly detailed level-of-inspection, and ultimately arriving at complete disassembly for individual part inspection and (as required) replacement.

4.1 IN THE BEGINNING

These flight-time based inspections brought about the first requirement for skilled planning and scheduling. Because of the practice of swapping out aircraft engines from planes with heavily damaged fuselages, it was a routine occurrence for the 120-hour inspections of the airframe and engine on one aircraft to come due at different times. This meant that aircraft were taken off-line twice as often as they would have been with coincident airframe and engine 120-hour inspections. Additionally, because the military had this tendency to fly all their aircraft on every mission, most aircraft would come due for their major inspections simultaneously. This, of course, would result in the entire squadron being "off-line" rather than just one or two aircrafts.

The flight-hour-based inspections required some planning skill to keep airframe and engine inspections synchronized and even more skill at maintaining staggered off-line schedules for squadron aircraft. The engine swapping practice had to be modified to attempt to keep like inspection intervals of mated airframes and engines. Similarly, it was often necessary to "fly aircraft into checks." This was the practice of forcing the hourly inspections to occur on each aircraft at their pre-planned times by either creating additional flight hours by assigning training flights or single aircraft sorties to pre-selected aircraft or reducing flight hours by "sitting-out" a mission. The goal was to avoid taking more than one aircraft off-line for a major inspection at a time. This manipulation of parts and flying time sometimes required extraordinary chart, record and time-line development and tracking in order to meet the "one aircraft at a time" goal. Soon, aircraft squadron maintenance officers were given special training in order to sharpen these skills. These were the first Maintenance Planning and Scheduling Workshops. A point worth making here is that the jobs of maintenance planning and maintenance scheduling, as they exist in modern manufacturing plants today, are at least 100-fold more complex than the job of the aircraft squadron maintenance officers of the 1930s and 1940s.

In the 1950s, the U.S. Navy was ordered to adapt the maintenance practices of the Air Force for shipboard use or develop their own system for effectively maintaining fleet readiness. The result was the 3M (Maintenance and Material Management) System for maintenance of shipboard systems and equipment. Maintenance Requirement Cards (MRCs) that provided procedural steps for performing periodic maintenance actions were developed at a headquarters level office along with maintenance requirement listings that identified all maintenance requirements, in frequency sequence, for individual shipboard equipment installations and system suites. At the ship level, updates for equipment maintenance plans were received on a quarterly basis and as equipment modifications were installed. Each shipboard division officer transcribed the periodic maintenance requirements onto large poster-board schedules, incorporating the ship's operating plan into each schedule. Various maintenance actions required the ship to be in port, at sea or in some other special configuration (e.g., full speed operations). Ship operating plans drove these schedules by either allowing or prohibiting certain requisite conditions for applicable maintenance requirements. In this role, the division officer was the closest yet in function to today's manufacturing plant maintenance planner. Today, although the Navy still refers to it as the 3M system, it has evolved tremendously. The basic document describing the system—the 3M Manual—is more than 400 pages in length and covers the use of computer scheduling, Reliability-Centered Maintenance and similar updated maintenance methodologies.

A part of the Navy's 3M System was the correlation of maintenance and repair actions to one of three levels of the ship repair hierarchy:

1. Operational (shipboard level);
2. Intermediate (in port at a repair command or alongside a repair ship or tender) and
3. Depot (shipyard level repairs, modifications and ship overhauls).

The two additional levels, Intermediate and Depot, were scheduled for a duration determined by the number and size of the work packages designated for accomplishment. The objective was to make the availabilities at these two repair levels only as long as was absolutely necessary to complete the work, so the ship could be deployed again as quickly as possible. This kind of "tight as possible" scheduling eventually gave rise to the application of Critical Path Method (CPM) and Project Evaluation and Review Technique (PERT) for scheduling and project management methodologies (see Chapter 8 for a description of the history and use of these methods. Appendix D also provides a brief tutorial on constructing and analyzing CPM Schedules). Both methods for scheduling and managing maintenance activities are used today in the manufacturing industry.

4.2 DEVELOPING STANDARD PRACTICES

Up to this point, the various historical predecessors to modern Maintenance Planners performed in a variety of roles with a variety of functions and responsibilities. In today's manufacturing environment, the requirements for maintenance planners have become much broader in scope and more complex in execution. This situation has often led to modification of Planning and Scheduling responsibilities to parallel the skill level of the individual assigned as the planner. The entire maintenance operation suffered as a result. There is a fundamental process for maintenance planning and scheduling, and depending on the unique variables at every plant or facility, additional process requirements may be assigned to the planning and scheduling function. However, dropping any of the fundamentals would be detriment of the overall maintenance function, as would the assignment of additional requirements that could supersede or detract from the primary planning and scheduling function.

4.2.1 Basic Process

There are specific goals and objectives associated with maintenance planning and scheduling (see Chapter 3). Achieving them defines a basic planning and scheduling process. In the following paragraphs, the goals and objectives of the Maintenance Planner/Scheduler are used to derive a basic planning and a basic scheduling process. These are defined as *Minimum Standard Practices of the Maintenance Planner/Scheduler*. As stated previously, of all Maintenance Organization's activities, Planning and Scheduling have the most profound effect on timely and effective accomplishment of maintenance work. In order to achieve the level of Planning and Scheduling performance necessary for a truly effective maintenance operation and to ensure consistency of performance across time and people, the development of a Maintenance Department Standard Operating Procedure (SOP) for the Maintenance Planning and Maintenance Scheduling process is required (see Appendix C for a Planning and Scheduling–SOP template). Sound Planning and Scheduling practices are the starting point for an effective maintenance operation.

Only well-defined maintenance tasks can be planned.
Only well-planned work packages can be scheduled.

The basic goal of the Maintenance Planner is the avoidance of delay, which defines the planner's position and role.

Position and Role: The role of the Maintenance Planner/Scheduler (P/S) is to improve work force productivity and work quality by anticipating and eliminating potential delays through planning and coordination of personnel, parts and material and equipment access. The P/S is a pivotal position in the maintenance operation and who maintains a continuous dialogue between Operations and Maintenance. The P/S is responsible for planning, scheduling and coordination of all maintenance work (that can be planned) performed on the plant site.

In order to achieve the six primary objectives of the Maintenance P/S, the planner's responsibilities and duties require a thorough knowledge of the maintenance operation as well as several additional skills. Maintenance Planning and scheduling is the hub from which all maintenance activity, that can be planned is coordinated. Planning and Scheduling are the processes for defining how and when a job is to be performed and what resources will be required. It involves a broad spectrum of activity.

The P/S must know the job well enough that he or she can describe what is to be accomplished and can estimate how many labor-hours will be required. If the P/S does not know the requirements, the assigned crew will not know their expectations. In performance of their duties, the P/S is:

- The representative of the Maintenance Manager and principal contact and liaison path between Maintenance and Operations as well as other supporting and supported departments. In this capacity, the P/S ensures that all internal customers of Maintenance receive timely, effi-

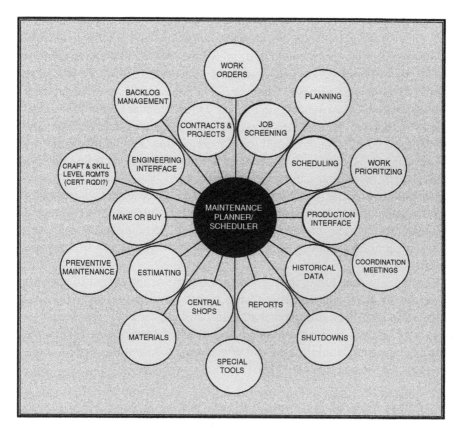

Figure 4-1 Planner/Scheduler Interfaces, Activities and Responsibilities

cient and quality service. He also maintains a continuing interest in the internal customer's needs. The P/S has a keen awareness of the customer's situation (schedule, problems, etc.), and therefore is able to help Operations balance their need for daily output with their need for equipment reliability through proactive maintenance.

- Responsible for long-range as well as short-range planning. Long-range planning involves the regular analysis of backlog relative to available resources. These two basic variables must be kept in balance if a proactive maintenance environment is to be established and sustained. The P/S is also responsible for maintaining records and files essential for meaningful analysis and reporting of maintenance related matters.

The primary activities of work planning are the identification of all technical and administrative requirements for a work activity and provision of the materials, tools and support activities needed to perform the work.

These items should be provided to tradespersons in an easy-to-use, complete work package. Effective planning should help ensure that consistent quality maintenance activities are conducted safely and correctly. Work planning is a finite, well-defined process (well defined by a Maintenance Department Standard Operating Procedure) that should be periodically assessed through field observation of work being performed. Direct feedback from the tradespersons (either via or in company with their immediate supervisor) to the planners should be a fundamental tool for improving the planning function and ensuring its adequacy. Thus, the "basic" maintenance planning process consists of three steps as shown in Figure 4-2.

Scheduling of corrective and preventive maintenance and of planned and forced outage work is necessary to ensure that maintenance is conducted efficiently and within prescribed time limits. Scheduling daily activities based on accurate planning estimates should improve the use of time-on-the-job.

Maintenance scheduling during planned outages is important to support the return of the plant to service on schedule (and within the approved budget) and results in improved availability and capacity factors. A contingency schedule should be maintained so that if a forced outage occurs, the forced outage time is minimized and effectively used, and all needed maintenance is performed prior to restart.

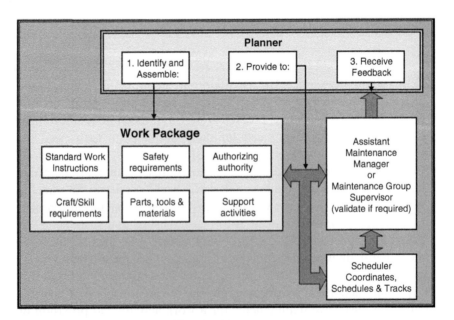

Figure 4-2 Basic Maintenance Planning Process

An effective schedule should assist management in controlling and directing maintenance activities and should enhance the ability to assess progress. The schedule should reflect the long-range plan and day-to-day activities. Effective scheduling should enhance the efficient use of resources significantly by decreasing duplication of support work, decreasing maintenance technician idle time and ensuring completion of planned tasks. The schedule should be the road map for reaching plant maintenance goals.

Scheduling, as the execution phase for planned maintenance tasks, is an integral part of the overall preparation for maintenance activities and should be performed in close coordination with planning activities. The completed and fully integrated schedule should be based upon details such as work scope, importance to plant goals, prerequisites and interrelations, work location and resources and constraints identified and developed during the planning process.

Effective daily schedules are needed to implement the maintenance activity plans represented by the integrated schedule. Management should track and periodically assess performance according to the daily schedule. Effectiveness of the daily scheduling process during normal operation should be a good indicator of how effective the daily schedule may be during major outages.

The integrated schedule should be reviewed by those responsible for implementation. It should be accepted and widely used by personnel involved in maintenance activities. Preparation of contingency schedules should decrease the time necessary to respond to problems or unforeseen perturbations, if they occur, and increase the information available for decision-making.

The basic process (see Figure 4-3) for preparing maintenance schedules is as follows:

1. The P/S determines, by discussion with the Supervisor and referring to vacation charts, the resources available and the expected total working hours during the schedule week.
2. The P/S reviews work orders from all sources that are "Ready to Schedule" then from personal knowledge, or after discussions with the operations supervisor, and arranges the work orders into priority order.
3. The P/S lists each work order on the schedule form.
4. Moderate Weekly Schedule Coordination Meeting to achieve a consensus between equipment custodians and maintenance/engineering supervisors as to the most effective near-term deployment of available maintenance resources.
5. Prepare and publish the approved weekly schedule. The crew supervisor performs daily Scheduling to coordinate new high-priority work

orders with those already in the weekly plan. Always striving to optimize schedule compliance despite essential schedule "breakers."

6. Schedule Follow-up to determine the level of schedule compliance and reasons for completion shortfalls. This is a "constructive" step towards future improvement.

4.2.2 Manufacturing's Influence

The manufacturing industry has unquestionably exerted the most influence on the function of the Maintenance Planner. The very nature of manufacturing creates a regular plethora of obstacles to straightforward planning and scheduling of maintenance tasks. The emphasis on steady and continuous operation of production equipment is the most significant obstacle. Operations managers have always demonstrated a reluctance (and still do so) to take production equipment offline "merely to perform maintenance." Fortunately, Lean Thinking has taught production oriented managers that improved equipment reliability through Total Productive Maintenance (TPM) can actually increase the net online time for production line equip-

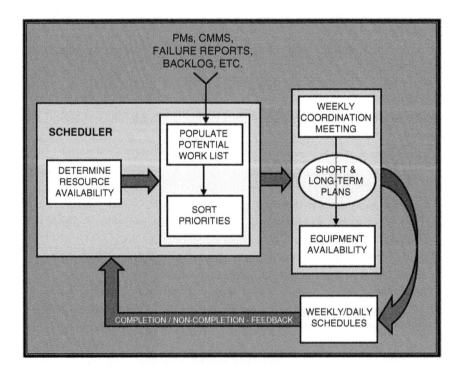

Figure 4-3 Basic Maintenance Scheduling Process

ment. Nonetheless, production schedules are still sacred cows and variations due to poorly planned maintenance activities are not well received.

Other characteristics of manufacturing operations in particular, lean manufacturing operations that tend to unsettle maintenance planning include:

- Pull System or Continuous Flow
- Interdependency of equipment in a production line
- Breakdowns
- Setup time
- Nonstaggered offline periods
- Dynamic production schedules

The most difficult situation that the Maintenance Planner faces is the potential for large variations in workload. For example, when all production equipment in a particular assembly or production line goes offline at a single time and within a short period of time, it produces the dilemma of maximum workload for a short amount of the total workweek and no workload for the remainder of the workweek. Yet a responsibility of the Maintenance Planner is creation of a steady workload for maintenance staffers.

4.2.2.1 Accommodating a Varying Workload

Accommodating (*Accommodating (1: to make fit, suitable, or congruous 2: to bring into agreement or concord: RECONCILE)*[3] a varying workload is only possible through the weekly schedule coordination meeting with operations. For example, a Planner must schedule maintenance work for six maintenance technicians for the following week. To simplify this example, assume that all technicians have the same skills and qualifications; the Planner must identify and schedule six × 40 hours, or 240 labor-hours of maintenance tasks. Operations has already provided the Planner with an equipment availability schedule for the following week. Line # 1 will be available, in its entirety, on Tuesday from 8:00 AM until 12:00 PM and Motor Control Center # 2 for Line # 3 will be offline on Thursday from 2:00 PM until 4:00 PM. The planner checks the "Ready Backlog" of work orders and sure enough, there are 160 labor-hours of work for Line # 1 and 80 labor-hours of work for Line # 3's MCC # 2—exactly 240 labor-hours. The only problem is that six employees just cannot produce 160 hours of labor in a four-hour period; they are pretty much tapped out at 24 labor-hours.

---

[3] *Merriam-Webster's 11th Collegiate Dictionary.*

That is where the Maintenance Schedule Coordination Meeting comes in. The Planner comes into the meeting armed with his preliminary "ready to schedule" or "available backlog" work list and operation's equipment availability schedule and proceeds to negotiate for additional equipment availability windows. Figure 4-4 shows the work orders scheduled for the operation-provided availability schedule in the darkened blocks. The availability windows that need to be negotiated are the un-shaded blocks outlined in bold with labor-hours filled in (only Monday and Tuesday are illustrated but the same process is applied for Wednesday through Friday). Of course, there may be overriding reasons that operations cannot take an indicated system/equipment offline, so the planner must have additional ready to schedule work in reserve to replace that work that is unable to be scheduled. Additionally, for example, operations may not be able to take Line # 2 Hydraulic system offline on Monday, but will be able to make it available Wednesday afternoon.

The weekly schedule coordination meeting is a give-and-take cooperative effort to achieve the best fit for maintenance work and operations' schedules. The planner must be armed with sufficient backup work to accommodate operations-driven perturbations to his preliminary schedule. A word of caution is necessary here. Planning is done prior to scheduling. If schedule coordination is an operations dictatorship such that a sizable portion of planned work is deemed "not possible" by operations, then the planner must *plan* for much more work than can be accomplished, simply as a contingency pool of backup work. Work planning is a detailed, time consuming and labor-intensive effort. It is done for work that *needs* to be performed. Planned work should not be considered as a frivolous effort that can be discarded out-of-hand either by operations or by the scheduler. Every effort should be made to stick to the preliminary schedule presented at the schedule coordination meeting. The fact that the scheduler should be armed with sufficient backup work to accommodate operations-induced schedule perturbations means that the scheduler should have one or two jobs in backup in the unlikely event that it is *truly impossible* to take a piece of equipment offline for maintenance.

Not all maintenance work requires equipment to be shut down or taken offline. Much of the weekly schedule will involve maintenance on running machinery; operating tests and measurements, inspection, predictive maintenance (PdM) and sampling are a few examples. Even though production equipment may not be required to be taken offline for this kind of maintenance work, coordination and finite scheduling is still a requirement.

In summary, it is important to reemphasize that the processes and examples just covered were the *basics*. They were intended to provide a concise overview of the Maintenance Planning and Maintenance Scheduling functions. There is, in fact, much more involved in both functions, lest you become too complacent and decide that the Maintenance

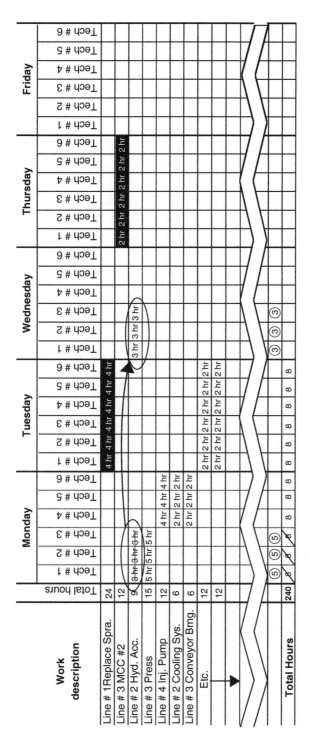

Figure 4-4 Negotiating Maintenance "Windows"

Planning and Scheduling effort is considerably easier than you thought. Chapters 6 and 7 will drill down to the details and provide the comprehensive processes and considerations that constitute Planning and Scheduling respectively.

4.2.2.2 Resources, Resources, Resources

In general, there seldom exists a scarcity of work to be planned and scheduled. The scarcities are normally associated with resources such as labor, material, technical documentation, procedural documentation, etc. Within resources, the most common and critical shortfall is labor. Before work can be either planned or scheduled, the maintenance P/S must know what resources are available to execute the work. The most effective method for the P/S to gain accurate information regarding labor resources is through the use of a weekly Labor Resource Report (LRR). The Labor Resource Report (see Appendix C for a sample LRR Form) should be submitted to the planner by each maintenance crew/team supervisor on Monday of the week prior to the period covered by the report. The LRR is much more than just a vacation schedule. Everyone knows that, even in a Lean Operating environment, there are many more demands on an employee's time than just work. Breaks, committees, meetings, training and a host of other things will detract from an employee's time available to work. Supervisors must account for all of these detractors, and specific time periods where applicable, in their weekly LRR to the maintenance P/S.

Additional resources required for efficient maintenance planning fall into several categories:

Material Libraries filed by each unit of equipment on which maintenance work is performed. The material library supports development of the bill of materials (BOMs) and the material cost estimate.

Labor Libraries filed in some conveniently retrievable manner—normally by unit of equipment, specific type of skill required, or by job code. The labor library supports development of job step sequence and the labor cost estimate.

Note: The content, development and use of these specialized planning "libraries" will be covered in detail in Chapter 7.

Equipment Records recording all pertinent data for equipment such as installation data, make and model, serial number, manufacturing, capacity, etc. Equipment Records should not be confused with Equipment History of actual repairs made to the equipment.

Prints, Drawings and Sketches as installed.

Purchasing/Stores Catalogs provide pertinent information not captured in the material libraries. Even where a materials library is in place, some form of stores catalog or vendor catalogs are used to develop it.

Standard Operating Procedures can be included in planning packages without repetitive documentation effort. Such procedures include safety, lockout, troubleshooting sequences, etc.

Labor Estimating System provides basic data for building job estimates. Even where a labor library is in place, some form of estimating system is used to develop it.

Planning Package File of previously developed packages for recurring jobs enables repeat usage with minimal repeat planner effort.

Additional Resources

1. Catalogs
2. Procedural Files
3. Service Manuals/Vendor Maintenance Bulletins
4. Equipment Parts Lists
5. Storeroom Catalogs
6. Estimating Manual(s)
7. Engineering Files
8. Equipment History

4.2.3 Appearance of Balance

Work planned and scheduled on the integrated weekly schedule should achieve a balance of maintenance activities in order to keep the various maintenance groups working at a steady pace and without long idle periods. At the same time, as pointed out earlier, the scheduling function is also meant to help Operations balance their need for daily output with their need for equipment reliability through proactive maintenance. A final balancing effort that the scheduler needs to perform is managing the backlog of work at a pre-determined level, while allowing for unplanned and nonrecurring mainte-nance activities. Balance must be achieved for maintenance resources to respond to breakdowns that might occur during the planned work schedule period and other activities such as special maintenance engineering investi-gations, investigation of out-of-range predictive maintenance results and a multitude of other irregular frequency maintenance activities. This all may sound like a balancing act worthy of a circus performer; in fact, it is a balancing act requiring considerably more skills than even the most talented circus performer might possess.

Achieving this balance requires adherence to some time-proven scheduling considerations, principles and procedures such as:

1. Lead time—needed work must be identified as far in advance as possible so that the backlog of work is known and jobs can be effectively planned prior to scheduling.
2. Work Backlog must be kept within a reasonable range. Backlog below minimum does not provide a sufficient volume of work to accommodate smooth scheduling. Backlog above maximum turns so slowly that it is impossible to meet customer needs on a timely basis.
3. Special or heavy demands cannot be scheduled unless backlog is addressed by providing additional resources or by relaxing priorities.
4. Jobs cannot be scheduled until all needs (parts, materials, tools, special equipment, the item to be worked, any special support) are available in the quantity required and at the time necessary.
5. Each available maintenance technician must be scheduled for a full day of productive work each and every day.
6. Emergency work may need to be done at the expense of scheduled jobs. The displaced scheduled jobs would constitute an overloaded schedule and result in work being carried over to the next schedule period unless addressed by a temporary increase in capacity, i.e., overtime.
7. Additional work (amounting to 10–15% of available scheduled manpower), should be identified and listed as fill-in for situations where scheduled jobs cannot be performed for a legitimate reason or other scheduled jobs have been completed in less time than planned. This 10–15% is not included in the schedule compliance calculation.

By adhering to these principles and considerations, the P/S ensures that:

- all maintenance needs are properly attended to;
- accurate evaluations are made as to the importance of each job with respect to the operation as a whole;
- customers have their work performed on a timely basis;
- equipment offline time experiences no, or minimum, delays;
- work is performed safely;
- overall maintenance cost is kept to a minimum.

Experience from the last 10 to 15 years has shown a relatively repeatable mix of maintenance work in a balanced backlog. The work mix is shown in Table 4-1. The table is provided for information purposes only and the numbers should not be considered as targets for the scheduling of planned work.

Table 4-1
Typical Balanced Backlog

Source of Planned Work	Percent
Results of PM inspections	30%
Scheduled component replacements	20%
Overhauls/rebuilds	15%
Internal Customer Input (Operators and Supervisors)	10%
Engineering project support	8%
Safety work	5%
Analysis of repair history	5%
Management directed work	4%
Service requests	2%
Accident damage	1%
Total Planned Work	**100%**

Every plant will have different influences on defining a balanced backlog of maintenance work. Local determination of what constitutes a balanced backlog of work should be made in order to recognize when the mix of work is out of balance, which indicates a process problem possibly requiring an investigation. When the considerations above are applied and the scheduling guidance of Chapter 7 is followed, a balanced backlog will result that contains whatever mix of work that is appropriate for your plant.

5

Organization Alternatives

In this chapter, various organizational styles in use for Maintenance Operations are discussed, with an emphasis on defining the strengths and weaknesses of each of these styles. The role of the Maintenance Planner/ Maintenance Scheduler and the execution of their responsibilities can be either enhanced or severely handicapped by the organizational structure of the Maintenance Department and by the Maintenance Department's organizational relationship with other functional areas of the plant. If you, or the Maintenance Operation within your plant, are undertaking an organizational makeover to recognize the Planning and Scheduling function as a designated position, the information in this chapter can be invaluable for ensuring success. Instituting the wrong organizational style or structure could end up making your Maintenance Operation less effective than it was before creating the Planner/Scheduler.

Before beginning the discussion of the various styles for organizing the maintenance function, it is relevant to define the corporate organizational configuration. This is necessary to illustrate just where the individual plant and the plant functions fit within the overall corporate structure. While there can be many subtle variations, most large corporations have traditionally been organized in the general form shown in Figure 5-1A and B. Please take note of the level 6 Maintenance Manager's responsibilities and then the entry "Equipment Maintenance" listed under the Production Manager's responsibilities. Up until 10 to 15 years ago, production equipment maintenance was exclusively an operations department function in virtually all manufacturing plants. While many plants today have separated out production equipment maintenance and assigned the function to another level 6 manager, the majority, by a slimmer margin every day, still retain that function under the Production Manager.

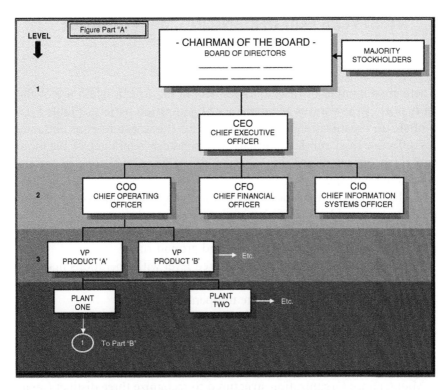

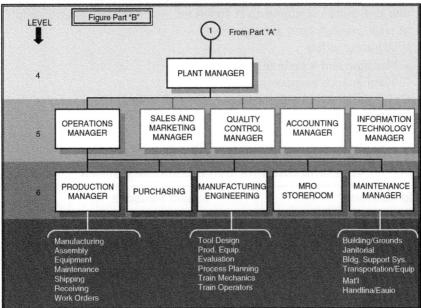

Figure 5-1 (A and B) The Generic Corporation

5.1 STYLES FOR ORGANIZING MAINTENANCE OPERATIONS

There are a number of driving forces behind the selection and use of the most typical of the various Maintenance Organization patterns (Table 5-1). Too often, an organizational style is chosen for the wrong reasons and may be less effective than another alternative. A breakdown of some of the more common reasons cited for using a particular organizational style or pattern includes the following.

The Maintenance Organization style used in any plant must be structured at a minimum to address the following issues effectively:

- Set forth organizational principles and ground rules.
  - ○ Maintenance management structured level with operations management
  - ○ Maintenance not subordinate to operations
  - ○ Supportive service to operations vs. subordinate function.
- Defined roles, responsibilities (including shared) and authorities.
  - ○ Operations department
  - ○ Maintenance department.
- Maintenance Organization structured to recognize three distinct (separate but mutually supportive) functions, so that each basic function receives primary level attention.
  - ○ Work execution
  - ○ Planning and scheduling
  - ○ Maintenance Engineering.
- Central and area assignments (where used) balanced to extent of economic soundness.

Table 5-1
Reasons Cited for Organizational Pattern

Valid Reasons	Invalid Reasons
Plant function	Pattern used at manager's previous plant
Plant size	Pattern used in other plant's departments
Plant layout	We have always done it this way
Maintenance style–TPM, Fixed Frequency, Fix when Fail (*reactive*)	CEO (top management) wants it this way (*without reason*)

- (When only one component of any organization optimizes, the organization as a whole suboptimizes).
- Planning and scheduling is one of the "prime legs" of the organization.
- Application of technical skills and knowledge.
- Nature of maintenance work and its control.
  - Routine
  - Emergency
  - Backlog relief.
- Impact of technological advancements on the nature of maintenance and production assignments.
- Organization for future.
- Encompasses job fulfillment.
- Rationalize the maintenance shift schedule.
  - Off shift schedule
  - Primary maintenance shift
  - Split shift need.

5.1.1 Style Variations

Seven definable variations in style are seen in the organization of maintenance activities:

1. Organization by *trade*;
2. Organization by *area*;
3. Organization *within production* department;
4. Joint *trade and area* organization structure;
5. *Contract* maintenance–partial or total;
6. Organization by *work-type*;
7. *Combination Styled Organization* combining work-type with area teams comprised of both maintenance and production personnel.

5.1.1.1 Trade Organization

Specialized trades whose unions enforce trade rules resulting in the organization of maintenance labor by trades is one of the dominant maintenance organization styles in use today (Table 5-2). Typically, maintenance jobs are first broken down into trade elements (electrical, instrument, sheet metal, machinist, welder, pipe fitter, etc.) and then each element is assigned to the appropriate trade group under the supervision of a trade supervisor, who directs the work of one or two trades covering the entire

Table 5-2
Trade Organization

Advantages	Disadvantages
1. Sufficient personnel are available to handle the work requirements of the plant.	1. Personnel are scattered around the plant and not closely supervised.
2. Considerable flexibility is available in assigning personnel of different trades to the various jobs	2. Time is lost in traveling to jobs
3. The total number of personnel can be held reasonably level, minimizing hiring and lay-offs.	3. Different personnel assigned to equipment, no one becomes proficient in its repair.
4. Specialists (electrical and instrument) are utilized more efficiently.	4. Interval between initial job request and completion for routine work can be longer.
5. Special maintenance equipment is used effectively	5. No one supervisor is responsible for total job completion, housekeeping or accountability.
6. One individual is responsible for all maintenance by skill or trade	
7. Accounting for all maintenance costs is centralized.	

facility. Thus, trade-styled organizations are *centralized* – each trade group will have work assigned throughout the facility by central scheduling or dispatching service.

Variation of Trade Organization: Functional Trade Assignments: Some multisite/multiplant companies use a variation of the trade pattern of organization—the functional-trade pattern. In this structural variation, each supervisor is assigned a major, functionally defined responsibility (e.g., maintaining electrical equipment or buildings and grounds) and provided with a work force composed of the requisite mix of trades. For example, the

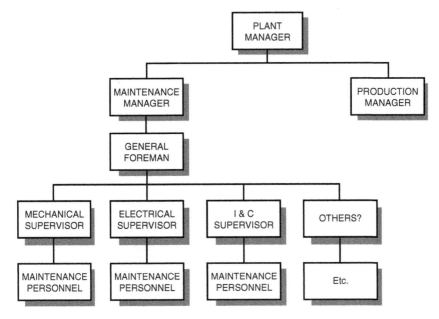

Figure 5-2 Trade Organization

buildings and grounds unit includes millwrights, painters, masons, gardeners and carpenters. The functional trade form of organization is based on trade-skill areas. It recognizes the functional tasks of organizing and administering maintenance work.

Functional work does not lend itself to the area type of supervision, either because it requires specialized skills or because the nature of the work requires maximum mobility. The plant engineer will assign such work to one of the central trade supervisors.

5.1.1.2 Area Organization

The area concept of supervising and controlling the maintenance function derives its name from the practice of designating relatively small maintenance areas in which the activities of assigned maintenance personnel are directed and controlled by one individual known as the area supervisor for maintenance (Table 5-3).

The area maintenance organization decentralizes the maintenance function. Maintenance crews are scheduled or assigned to areas within the plant, building or group of plants or buildings. Each area supervisor is responsible for maintaining uninterrupted production in his area. The personnel and

Table 5-3
Area Organization

Advantages	Disadvantages
1. Maintenance personnel are readily accessible to operations	1. The tendency exists to overstaff the area
2. Time spent traveling to a job is reduced	2. Major repairs are difficult to handle
3. Time lag is minimized between work request and work completion	3. There are more personnel problems and regulations pertaining to transfer, hiring, working overtime
4. Maintenance supervisors and personnel become better acquainted with the equipment and its spare parts requirements	4. Special equipment is difficult to justify because usage may be limited
5. Maintenance personnel are more closely supervised	5. Duplication of equipment occurs in the area maintenance shops
6. Production line or process changeovers is faster	6. More clerical help is needed if the area groups are large
7. There is greater continuity from one shift to another	7. Specialists are difficult to utilize effectively
8. Maintenance supervisors and personnel become more familiar with production schedules, problems, special jobs, etc.	

trade groups assigned have the required skill varieties to carry the normal workload within their area. When, and if, additional tradespersons are required, the area supervisor may requisition them from other area groups.

5.1.1.3 Production Department Maintenance

This is the older, historical structure seen before the evolution of maintenance as a separately managed discipline (Table 5-4). It is still found in

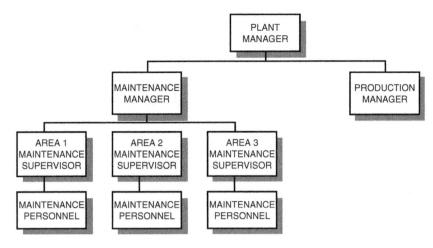

Figure 5-3 Area Maintenance Organization

many organizations, usually the smaller ones, that cannot economically justify separate supervision for maintenance crews of less than six or so tradesperson.

Production Department maintenance crews are responsible only for the production equipment within their production zone. Support services and systems, such as facility HVAC systems, centralized hydraulic and pneumatic systems, are maintained by a central maintenance group (or contracted out) that is directly responsible to Plant Operations.

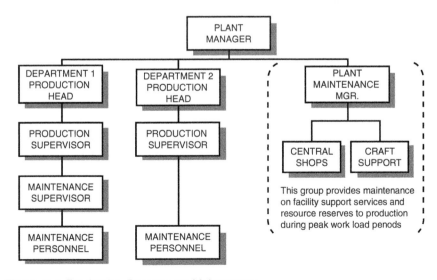

Figure 5-4 Production Department Maintenance

Table 5-4
Production Department Maintenance

Advantages	Disadvantages
1. Maintenance personnel are readily accessible to operations	1. Operations supervisors not qualified to direct a maintenance job
2. Time spent traveling to a job is reduced	2. Operations supervisors cannot give technical assistance to a mechanic
3. Time lag is minimized between work request and work completion	3. Operations supervisor may neglect maintenance in order to meet schedules.
4. Maintenance supervisors and personnel become better acquainted with the equipment and its spare parts requirements	4. The maintenance responsibility of the plant is divided.
5. Maintenance personnel are more closely supervised	5. The plant maintenance costs are harder to isolate and control
6. Production line or process changeovers is faster	6. Personnel problems are more pronounced than with area maintenance
7. There is greater continuity from one shift to another	7. The tendency exists to overstaff the area
8. Maintenance supervisors and personnel become more familiar with production schedules, problems, special jobs, etc.	8. Major repairs are difficult to handle
	9. There are more personnel problems and regulations pertaining to transfer, hiring, working overtime
	10. Special equipment is difficult to justify because usage may be limited
	11. Duplication of equipment occurs in the area maintenance shops

Table 5-4
Continued

Advantages	Disadvantages
	12. More clerical help is needed if the area groups are large
	13. Specialists are difficult to utilize effectively

5.1.1.4 Joint Trade and Area Organization

Under this type of organization, the central shop is expanded by subdividing the shop into a series of specialized trades, thus increasing the total number of tradespersons. Some tradespersons are permanently assigned to shops and others to areas of the plant to take care of minor repairs, adjustments, and even construction work, so that production can continue without interruption. Some trade activity (pump and turbine overhauls, re-tubing, rigging, machine shop work on lathes and grinders, valve repair, etc.) is centralized as a shop or central trade function and work is performed out in the field under the central trade function's supervision–not area supervision. (Electrical and instrument work are also good examples of functional work.) The central maintenance shop supervisor(s) is responsible for shop work, crew administration, and field consultation, as required for meeting production demands.

Special project work is under the direct supervision of the *special projects supervisor*. The types of work designated as special projects are varied. The special projects supervisor may be assigned as a relief supervisor for another supervisor, who has an overload of work or has been assigned to execute some major or special project. The assignment may be either on a temporary or full-time basis depending on plant size and special project work volume.

All fieldwork, except functional work and specifically assigned special project work, is under the direct supervision of an *area supervisor*. Within a given plant area, the area supervisor directs the activities of all tradespersons assigned to perform work except those reporting to a functional tradesperson. The area maintenance supervisor is responsible for obtaining materials, special tools, and equipment indeterminable by planning and other needs that will expedite the work in progress. Timekeeping and the distribution of hours to jobs for payroll, planning, and accounting purposes also are delegated to the area maintenance supervisor.

The area maintenance supervisor is not held responsible for supervision of personnel performing functional work in this area, or working under a supervisor responsible for a special project affecting the area. However, the area supervisor is responsible for bringing to the attention of the assigned functional or special project supervisor any instances observed of inadequate or inappropriate methods, poor quality, improper conduct or behavior that is detrimental to the area.

All of the work must be coordinated by planning with the field area supervisor, trade or functional supervisor, and operations to assure timely compliance.

Management and plant engineers, aware of the difficulty in balancing service and maintenance costs, have attempted to resolve this problem by combining a central group with an area or departmental group. Combinations of the basic plans are widely used in the industry. The variations and modifications are numerous (Table 5-5).

5.1.1.5 Contract Maintenance – Partial or Total

Within this organizational structure, maintenance work is left entirely up to contracted maintenance forces. The plant may choose to keep certain trades and turn all other work over to the contractor for supervision and work execution. The plant may elect to plan and schedule the contractor force or let them handle the P/S. In normal practice, outside contractors are employed to perform: (1) backlog work; (2) recurring nonemergency work; and (3) designated peak load work during situations such as shutdowns, turnarounds, construction, jobs, etc. The decision to use outside forces is usually based on a tradeoff study of cost and a number of other, less tangible factors.

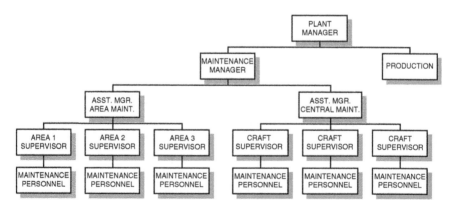

Figure 5-5 Joint Trade and Area Organization

Table 5-5
Joint Trade and Area Organization

Advantages	Disadvantages
1. A group of central technicians capable of handling the large projects and major repairs throughout the plant	1. Central technicians assigned to work throughout the plant with resultant high travel time and less job s upervision
2. Good control of maintenance costs	2. Tendency to go to fixed crews with area preference
3. Area technicians available to support production centers	3. Tendency to overstaff an area
4. Area technicians familiar with key equipment in the production centers	4. Duplication of equipment
5. Quick response time	5. Skill levels must be balanced properly

For a variety of reasons, management sometimes chooses to contract plant maintenance to an outside firm. In the case of large, multisite and multiplant companies, a single firm often has the contract for maintenance at all of the company's plants. In other cases, the maintenance contractor is often an offshoot division of the prime construction contractor that built the

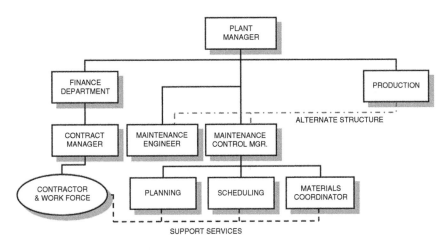

Figure 5-6 Contracted (Outsourced) Maintenance Organization

Table 5-6
Contracted (Outsourced) Maintenance

Advantages	Disadvantages
1. Contractor theoretically has greater ability to flex staffing with workload	1. Difficult to communicate job details to workers
2. Specific tasks can be targeted	2. Contractor shares no ownership to the equipment being maintained
3. Specific trade skills do not need to be staffed	3. Also used in high labor rate areas where heavy construction makes retention of tradespersons difficult and only possible at rates, which destroy the balance of the wage structure.
4. Often used where operating personnel are salaried, non-union technicians	
5. Quite common in refineries	

facility. In still other cases, only a portion of the work is contracted. Note in Figure 5-6 that an alternative organizational structure places the Maintenance Control Manager and Maintenance Engineer in the Operations leg of the structure. In either case, the maintenance functions do not operate at the level of other plant departments, even though they may report directly to the Plant Manager (Table 5-6).

5.1.1.6 Organization by Work Type

As stated earlier, in an effective maintenance organization, there must be recognition of, and provision for, the three types of maintenance work: reliable *routine* maintenance services, timely backlog relief, and prompt emergency response.

Routine maintenance includes preventive maintenance activities such as lubrication, visual inspection, and testing. It also includes predictive maintenance activities such as obtaining vibration measurements and drawing lubricant samples for analysis. In short, routine maintenance is comprised of

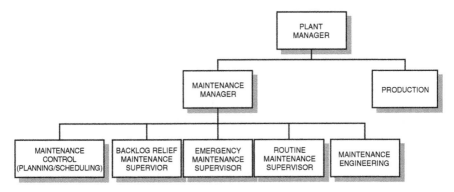

Figure 5-7 Organization by Work Type

nearly all of what is referred to as Preventive/Predictive Maintenance (PM). There may be both preventive and predictive routines that are complex enough to be performed by backlog relief personnel as planned work packages, or by Reliability Engineering for predictive routines requiring special skills or training.

Backlog relief deals with the bulk of the maintenance department work: the investigation, repair and restoration activities, the need for which is identified by operators or by personnel performing routine or PM/PDM actions.

Emergency response requires immediate action to address breakdowns and other suddenly developing conditions. Traditionally, emergency response consumes excessive maintenance resources. Although an absolute necessity, emergency response is a totally reactive function. Proper attention to the routine work and timely backlog relief will minimize the demand for emergency response.

The work-type organizational structure is consistent with the natural manner in which the function must be controlled. Effective control of work-type maintenance depends upon clear accountability for each type of demand placed upon the organization. An organization can be structured to facilitate control of each type of work. Such a structure is composed of three major operating groups covering the three principal types of demand. The basic concept of this structure is the establishment of two minimally sized crews to meet the routine and emergency demands and a third, larger group devoted to planned backlog maintenance.

The routine/preventive maintenance group is responsible for the performance of all management-approved routine tasks in accordance with detailed schedules and established quality levels. Their work:

- is specifically defined;
- is performed according to a known schedule;

- is performed in a planned pattern;
- involves a consistent work content;
- has well-defined and predictable time requirements.

The group is not interrupted by emergencies or backlog, thereby protecting the integrity of the preventive maintenance schedule.

The emergency group, sometimes called the Do it Now (DIN) Squad, has the responsibility for handling essentially all emergency demands, using assistance only when necessary. It is not possible or economical to plan and schedule all maintenance activities. To handle emergency work a DIN Squad is established with 10–20% of the maintenance work force. These people are highly skilled, highly trained, self-motivated individuals. They are often equipped with special gear, such as:

- motorized cars;
- two-way radios;
- complete sets of standard tools;
- selected special tools;
- high usage, critical spares.

Assignments for these individuals are handled through a dispatcher, normally someone in the Planning and Scheduling Group, who logs calls (or Emergency Work Orders) including:

- Person requesting service
- Time request was made
- DIN Squad member(s) assigned to job
- Time work started and completed by DIN Squad.

Alternately, the dispatcher or scheduler receiving the Emergency Work Order may pass it directly to the DIN Group Maintenance Supervisor who assigns and tracks the work.

A backlog of low-priority work, usually requiring less than two hours for completing, is available for these employees to perform when no emergency or other DIN assignments are waiting. This allows the planned maintenance group to apply their manpower to backlog relief (Table 5-7).

5.1.1.7 Combination Styled Organization (Work-Type Combined with Teams Comprised of Maintenance and Production Personnel)

With the advent of advanced technology and computer process control, operating personnel are frequently in a standby or monitoring mode.

Table 5-7
Organization by Work Type

Advantages	Disadvantages
1. Clear accountability for each type of demand placed on the organization	1. Not easy to achieve in tough union environments.
2. Structured to cover the three principle type of maintenance work	2. Requires higher staffing level of multi-skilled personnel
3. Matched skills and personality traits to functions	3. Personnel tend to get locked to specific functions and tend to loose focus of the "Big Picture"

Advanced production technology also often requires extensive process knowledge on the part of maintenance personnel in order to troubleshoot problems and failures. Together, these, and additional factors born of Total Productive Maintenance (TPM), have fostered the creation of the combination style organization structure. When coupled with self-empowered teams in the Lean Maintenance Environment, the diversified set of available skills provide unique approaches to problem-solving as well as process and reliability improvement. Combination or multiskilled, empowered teams are delegated broad responsibilities for operation and maintenance of specific facility functions, equipment/systems, areas, or processes.

The influence of TPM in this style of organization is to focus on equipment management, not simply on maintenance. Equipment Management (EM) is pursued through autonomous small group (team) activity comprised of operators, engineers, and maintenance personnel. The team's pursuit of improvement in equipment reliability, and in process and quality improvement, never stops.

The aim of team efforts is to optimize overall equipment effectiveness, optimize safety, and eliminate breakdowns by using a thorough system of maintenance throughout equipment's entire life span. Through involvement, operators develop stewardship of, and an affinity for "their equipment." Team members involved in problem-solving develop a stronger desire to see the problem fixed. Therefore the entire team participates in EM activities to ensure that the problem does not recur. The empowered team structure fosters pride in their joint accomplishments. The team organization trains, solves problems, and performs its work as a self-contained entity. Peer support, as well as peer pressure to maintain high levels of involvement, is important within each team, but at the same time, interteam competition is healthy and encouraged (Table 5-8).

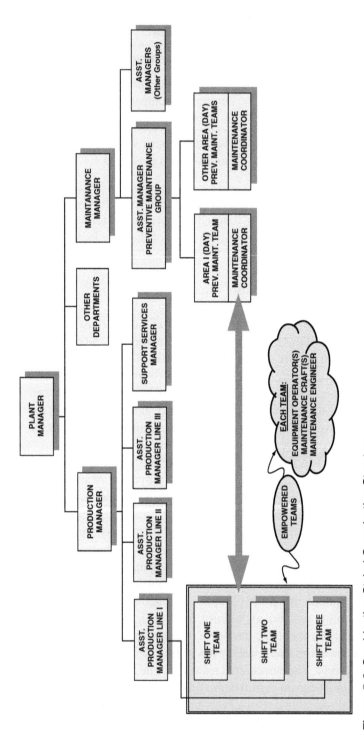

Figure 5-8 Combination Styled Organization Structure

Table 5-8
Combination Styled Organization

Advantages	Disadvantages
1. Teams trained to recognize and correct basic day-to-day problems	1. Fosters emergency response only
2. Participation, self-motivation and team responsibility strongly emphasized	2. Maintenance troubleshooting requires extensive process knowledge
3. Strong motivation toward training and versatility	3. Maintenance neglected in favor of meeting the production schedule
4. Generates higher motivation and individual satisfaction	
5. Improved joint understanding and dialogue	
6. Retains best elements of the combined craft-area structure	
7. Retains best elements of the comb	

A problem with depicting organizational attributes with the standard organization "tree" charts is the fact that they do not effectively convey support, liaison, and communication paths. The problem becomes even more pronounced when the Planner/Scheduler functions are introduced into the organization's structure. There are a number of ways to incorporate planning into any given style of organization, not all of which are appropriate or effective.

5.1.2 Total Productive Maintenance

TPM is a company-wide equipment maintenance system involving all employees, from top-level management to production line workers and the building custodians. It is just one of many approaches to maintenance. The word "total" in TPM is common to:

Total Equipment Effectiveness in pursuit of profitability, not simply maintenance cost reduction.

Total Maintenance System including Maintenance Prevention (MP), Maintainability Improvement (MI), and good old-fashioned Preventive Maintenance (PM), a *maintenance plan* for the equipment's entire life span. MP translates into maintenance-free design; it is pursued during the equipment design stage. MI translates into repairing or modifying equipment to prevent breakdown and to facilitate ease of maintenance.

Total participation of all employees: The idea behind TPM is not revolutionary. Simply stated, it is cooperation to get the important job of maintenance done reliably and effectively! It is supplementary to, rather than a replacement of, established principles of successful maintenance management. Built around six focal points, TPM combines concepts of continual improvement, total quality, and employee involvement:

- Activities to optimize overall equipment effectiveness.
- Elimination of breakdowns through a thorough system of maintenance throughout the equipment's entire life span.
- Autonomous operator maintenance (this does *not* imply that operators perform *all* maintenance).
  - Use lower-skilled personnel to perform routine jobs that do not require skilled tradespersons.
  - Use operators to perform specific routine maintenance tasks on their equipment.
  - Use operators to assist technicians in the repair of equipment when it is down.
  - Use computerized technology-enabling operators to calibrate selected instrumentation.
  - Use technicians to assist operators during shutdown and start-up.
- Day-to-day maintenance activities involving the total work force (engineering, operations, custodians, maintenance, management).
- Company-directed and motivated, yet autonomous small group activities. Small group goals to coincide with company goals.
- Continuous training
  - Formal
  - On-Job Training (OJT).
  - One-point lessons
  - Team members train each other.

While the dual goals of TPM are zero breakdowns and zero defects, TPM is more about performance improvement, employee interaction, and positive reinforcement than it is about maintenance of specific technology. Human resource factors and technical factors must be balanced.

TPM works toward elimination of six formidable obstacles to equipment effectiveness:

- Downtime
  - Equipment failure—from breakdown
  - Setup and adjustment.
- Speed Losses
  - Idling and minor stoppages—due to abnormal operation of sensors, blockages of chutes, etc.
  - Reduced speed—due to discrepancies between designed and actual speed of equipment.
- Defects
  - Process defects—scrap, downgrades, rejects, returns, etc.
  - Reduced yields—from all resources: raw, packaging, energy, labor, etc.

Group or team activities are promoted throughout the organization to gain greater equipment effectiveness. Operators are trained to *share* with maintenance personnel the responsibility for routine maintenance. This is referred to as "autonomous maintenance." Routine maintenance normally includes:

- housekeeping;
- equipment cleaning;
- protection of components from dirt;
- lubrication by operators;
- equipment inspection by operators and by maintenance;
- set-ups and adjustments.

Autonomous Maintenance may or may not include minor equipment repairs. Each plant must decide how autonomous maintenance is to be defined. Operators who perform autonomous maintenance must first receive significant skills training and then must be certified as they progress to higher skill levels. As operators become trained, an organized transfer of tasks takes place.

The TPM approach does not preclude need of an integrated maintenance program including computerized support, formalized planning and scheduling, and insightful equipment history.

TPM is not a short-lived, problem-solving, maintenance cost reduction program. It is a process that changes corporate culture and permanently improves and maintains the overall effectiveness of equipment through active involvement of operators and all other members of the organization. The required TPM investment, as well as the return, is very high. Systematic TPM development and full implementation requires two to three years to complete. Over time, the cooperative effort creates job enrichment and pride, which dramatically increases productivity and quality, optimizes equipment life cycle cost, and broadens the base of every employee's knowledge and skill set. Because equipment experiencing frequent breakdowns does not lend itself to proactive maintenance, the company must bear the additional expense of restoring equipment to its proper condition and educating personnel about *their* equipment before implementing TPM.

At this juncture, it is once more important to reiterate that TPM is the required maintenance organization operating mode in the Lean Maintenance Environment.

5.1.3 Reliability-Centered Maintenance

One of the enhancements of TPM that "fine tune" it for the Lean Maintenance operating mode is the selective use of several Reliability-Centered Maintenance (RCM) features. In Chapter 2, it was stated that one of the objectives of lean maintenance was the elimination of unnecessary maintenance; another was maintenance optimization. This is where RCM is "selectively" applied to "fine tune" TPM. RCM is a primary tool applied by the Reliability Engineering group.

RCM was born out of a 1960 FAA/Airline Industry Reliability Program Study that was initiated to respond to rapidly increasing maintenance costs, poor availability and concern over the effectiveness of traditional time-based preventive maintenance. The study centered around challenging the traditional approach to scheduled maintenance programs, which were based on the concept that every item on a piece of complex equipment has a operating time limit at which complete overhaul is necessary to ensure safety and operating reliability.

The generally used definition of RCM is that it is a process used to determine the maintenance requirements of any physical asset in its operating context. Perhaps a more complete, or accurate, definition is that RCM consists of processes used to determine what must be done to ensure that any physical asset continues to do whatever its users want it to do in its present operating context. In strict context, RCM does not concern itself with issues such as labor efficiency or waste elimination. Therefore, if RCM is to be applied in a lean operation, these constraints must be enforced by the TPM process.

There has been a proliferation of maintenance programs calling themselves RCM. In an effort to curtail these "offshoots" of RCM, the Society of

Automotive Engineers has developed and issued SAE JA-1011, which provides some degree of standardization for the RCM process. The SAE standard defines the RCM process as asking seven basic questions from which a comprehensive maintenance approach can be defined.

- What are the functions and associated performance standards of the asset in its present operating context?
- In what ways can it fail to fulfill its functions?
- What causes each functional failure?
- What happens when each failure occurs?
- In what way does each failure matter?
- What can be done to predict or prevent each failure?
- What should be done if a suitable proactive task cannot be found?

The first four questions constitute a functional Failure Modes and Effects Analysis (FMEA) and the last two define the appropriate maintenance approach. The vitally important fifth question determines failure detectability (whether the failure is 'hidden' or 'evident') and whether and in what way safety, the environment or operations are affected. These seven questions can only sensibly be answered by people who know the asset best; this includes maintainers and operators, supplemented by representatives from OEMs. In a full-blown RCM analysis, this group is guided through each step by a competent "facilitator" who is an expert in the RCM process and its application rather than the system expert. In the Lean Maintenance Environment, "full-blown" RCM analysis is usually not necessary, with the possible exception of extreme cost or high potential for severe environmental or safety impact, hence the Lean Maintenance "selective RCM application."

Because the Lean Maintenance objectives to be examined are elimination of unnecessary maintenance and maintenance optimization, questions 6 and 7 are the ones of immediate interest. Maintenance analysis is performed to determine the answers. Note that the maintenance analysis process, as illustrated in Figure 5-9, has only four possible outcomes:

1. Perform interval (time- or cycle-) based actions, or
2. Perform condition-based actions, or
3. Perform no action and choose to repair following failure, or
4. Determine that no maintenance action will reduce the probability of failure AND that failure is not the chosen outcome (redesign or redundancy).

Additional responsibilities of Reliability Engineering in the Lean Maintenance Environment include the elimination of unnecessary

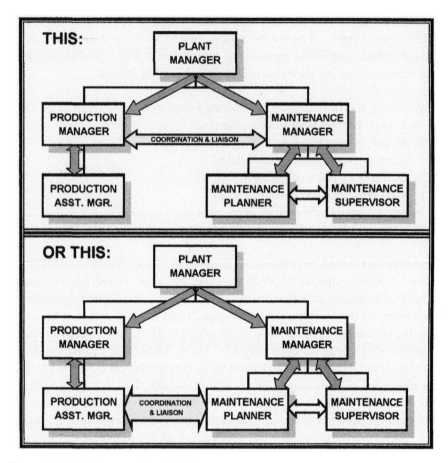

Figure 5-9 Maintenance Analysis Process

maintenance activity and maintenance optimization. The responsibilities of Reliability Engineering for the establishment and execution of maintenance optimization using Computerized Maintenance Management Systems (CMMS) Unscheduled and Emergency reports, Planned/Preventive Maintenance reports and Condition Monitoring (CM)/Predictive Maintenance (PdM) analysis include:

- Develop CM tests and PdM techniques that establish operating parameters relating to equipment performance and condition and gather data.
- Analyze CMMS reports of completed CdM/PdM/Corrective work orders to determine high-cost areas.
- Establish methodology for CMMS trending and analysis of all maintenance data to make recommendations for:

1. Changes to PM/CM/PdM frequencies;
2. Changes to Corrective Maintenance criteria;
3. Changes to Overhaul criteria/frequency;
4. Addition/deletion of PM/CM/PdM routines;
5. Establish assessment process to fine-tune the program;
6. Establish performance standards for each piece of equipment;
7. Adjust test and inspection frequencies based on equipment operating (history) experience;
8. Optimize test and inspection methods and introduce effective advanced test and inspection methods;
9. Conduct a periodic review of equipment on the CM/PdM program and delete the equipment no longer requiring CM/PdM;
10. Remove from, or add to, the CM/PdM program that equipment and other items as deemed appropriate;
11. Communicate problems and possible solutions to involved personnel;
12. Control the direction and cost of the CM/PdM program.

The joint FAA/Airline industry study performed in 1960, referred to earlier, revealed that many types of failures could not be prevented, or even reduced, by operating time limit overhauls no matter how extensively they were performed. Two unexpected findings from the 1960 FAA/Airline Industry Reliability Program Study were as follows:

1. Scheduled overhauls had little effect on the overall reliability of a complex item unless the item had a dominant failure mode.
2. There were many items found for which there was no effective form of scheduled maintenance.

The 1960 study and related studies performed after that initial program indicated that all components do not follow the traditionally accepted "operate reliably then wear out" failure probability. They tend to follow a variety of failure probabilities as illustrated in Figure 5-10.

The failure probability distribution must be considered when defining a PM/CM/PdM optimization strategy for equipment. For example, CM is likely to be the maintenance mode of choice for a component whose failure probability is "slow aging." Similarly, fixed frequency overhaul or replacement, or an appropriate PdM technique, is a likely choice for a component whose failure probability is "Best New."

The Reliability Engineer can employ any of several techniques for determining optimum frequencies for maintenance actions including statistical determination, Condition Monitoring (CM) and PdM Analyses, and analysis of equipment history records. The latter technique, analysis of

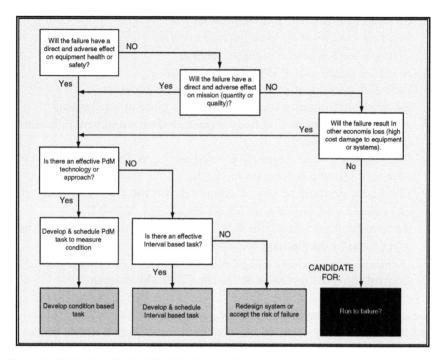

Figure 5-10 Failure Probability Distribution Patterns

equipment history records, is made vastly easier by effectively implemented CMMS or Enterprise Asset Management (EAM) systems. Ensuring the completeness and accuracy of the historical data archived in CMMS or EAM is under the purview of Maintenance Planning and Scheduling.

CM is differentiated from PdM primarily by the use of in-place instrumentation and in-place testing (self-test) capabilities to gather data indicative of component, equipment or system condition. In its simplest form, condition monitoring can employ procedural documentation that provides an area for recording readings of installed meters and gauges when they are pertinent to the preventive maintenance task being performed. Such data can provide a simple means for evaluating the effectiveness of the maintenance and optimizing scheduling. An example of the most basic level of CM and its use in maintenance optimization is illustrated in Figure 5-11.

A major responsibility of the reliability engineer is that of eliminating recurring failures. He accomplishes this through the application of Root Cause Failure Analysis (RCFA). Failures are seldom planned for and usually

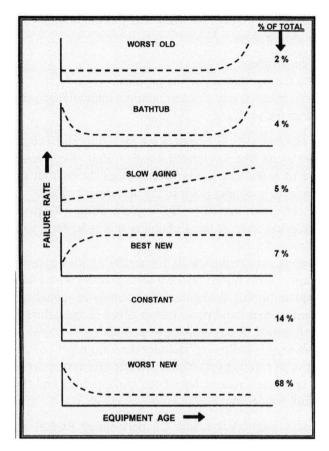

Figure 5-11 Application of CM to Maintenance Optimization

surprise both maintenance and production personnel and they nearly always result in lost production. Finding the root cause of a failure provides you with a solvable problem removing the mystery of why equipment failed. Once the root cause is identified, a "fix" can be developed and implemented.

There are five basic phases in performing a RCFA:

Phase I (data collection): It is important to begin the data collection phase of root cause analysis immediately following the occurrence identification to ensure that data are not lost. (Without compromising safety or recovery, data should be collected even during an occurrence.) The information that should be collected consists of conditions before, during, and after the failure occurrence; personnel involvement (including actions taken); environmental factors and other information having relevance to the occurrence.

Phase II (assessment): Any root cause analysis method may be used that includes the following steps:

1. Identify the problem.
2. Determine the significance of the problem.
3. Identify the causes (conditions or actions) immediately preceding and surrounding the problem.

Identify the reasons why the causes in the preceding step existed, working back to the root cause (the fundamental reason that, if corrected, will prevent recurrence of these and similar occurrences throughout the facility). This root cause is the stopping point in the assessment phase.

Phase III (corrective actions): By implementing effective corrective actions for each cause reduces the probability that a problem will recur and improves reliability and safety.

Phase IV (inform): Entering the RCFA results on the appropriate RCFA worksheet or report form is part of the inform process. Also included is discussion and explanation of the results of the analysis, including corrective actions, with management and personnel involved in the failure occurrence. RCFA reporting should be as complete as possible.

Phase V (follow-up): Follow-up includes determining if corrective action has been effective in resolving problems. An effectiveness review is essential to ensure that corrective actions have been implemented and that they are preventing failure recurrence.

There are many methods available for performing RCFA; selecting the right method for RCFA can speed the entire process up so that you can proceed to the "fix" stage more quickly. Some of the more commonly used RCFA methods include:

- Ishikawa method of cause and effect analysis (fishbone diagramming);
- Fault tree analysis
- Sequence of events analysis
- Events and causal factor analysis;
- Change analysis;
- Barrier analysis;
- Management oversight and risk tree (MORT);
- Human performance evaluation;
- Kepner–Tregoe problem solving and decision making.

5.1.3.1 Incorporating Work Planning

Incorporating work planning into the maintenance organization first requires pre-preparation to ensure the requisite tools and conditions are in place to facilitate the efforts of work planners. Correctly and efficiently performing the planning function requires management to provide adequate guidance on the level of control necessary to ensure consistent quality maintenance of plant equipment. The requirements to provide procedures for safety-related equipment and equipment important to plant safety should be clearly spelled out in plant standard operating procedures (SOPs), Safety Instructions and related plant policy documentation.

Large disparities exist throughout industry in the level of detail and procedural guidance provided to maintenance personnel for performing work on plant equipment. Many plants rely heavily on "presumed trade skills" but they have not assessed the actual skill levels of their personnel. For example, it is commonly accepted that an electrician possesses the necessary skills to install wiring lugs; however, the manufacturing industry as a whole, continues to have problems with loose wiring.

Presumed trade skills should be given careful consideration when preparing planned work packages, job requests, and work instructions to ensure that additional training, worker qualifications, or job oversight/quality control are included, if required. For example, work instructions for nonfacility contractors may need to include more detail, inspections, or supervisory guidance.

To reduce problems caused by inadequate instructions being provided to the tradesperson, managers should establish minimum levels of trade proficiency and implement training programs to ensure that the expected trade skill levels are developed and maintained. Maintenance Supervisors should be responsible for assessing skill levels and identifying training requirements. Deficiencies identified through daily activities, industry experience, or root cause analysis may result in the identification of additional training needs to maintain or upgrade this skill level. For work beyond expected skills, more detailed work instructions will need to be provided to the tradesperson.

"Presumed trade skills" are work skills that should be common knowledge (for the skill level designated) to the individual performing the work. The skill level of every maintenance employee should be assessed. Maintenance trades should then be formally trained to advance to successively higher skill levels. By means of a qualification and certification process tied to the training program, certification as qualified to perform at each skill level should be a formalized process. Unqualified maintenance workers should be assigned to work under the supervision of a qualified individual.

Without adequately defined and controlled trade skills, the planner's function is rendered at least 10-fold more complex and difficult. Given ill-defined trade skills, each work plan will need to be tailored to the individual assigned to the work, and specific individuals will need to be assigned to the work before the planner can complete the work plan instructions. Such a dichotomy is often referred to as a "self-eating watermelon."

Where the planning function "fits" within the maintenance operation and what mechanisms exist to enhance the planner's effectiveness should be well defined if the work planning function is to realize all of the beneficial effects that are possible. Work planning encompasses the coordination of various input resources (material, labor, and equipment), as required, to complete each job in an orderly manner and at least overall cost. Supervisors are relieved of much indirect activity, enabling them to spend their time in the plant, overseeing the trade crews while the planning function is performed by quasi-managerial personnel. Just how this will be accomplished requires decisions concerning

- where will the planning group fit into the overall organization?
- who will be responsible for establishing and administering the program?

Planning installations that fail often do so due to insufficient attention to these details. The importance of incorporating this vital function into the overall facility strategy is missed. The planning function should report through the maintenance staff at least one level above the first maintenance supervisor level supported on a day-to-day basis. If the planning function reports to the first line maintenance supervisor, there is a tendency to use the planner for work other than planning. The planner's job should be rated at least on a par with the first line supervisor, otherwise problems will arise regarding the planner's authority.

To gain the favorable attitude of both the craftsman and maintenance supervisors, active program support by plant and maintenance management is essential. By placing the planning group directly under the supervision of maintenance management and creating planner job positions equivalent in grade to first line supervision, functional importance is emphasized.

If the planning function is positioned too low in the organization, it does not receive proper management attention when decisions are required. Therefore, if there are more than four positions in the maintenance control organization (including planners, schedulers, material coordinators, clerks, and dispatchers), they should report through a maintenance control manager to bring coordination, functional discipline and integrity, as well as managerial acumen and clout to the maintenance services group. Such coordination is not always performed effectively by the maintenance manager who is consumed more by direction of first line supervision, budgetary control, and response to upper management. When the burden of formal planning and scheduling is added, it often becomes too much for the maintenance manager to address adequately.

A conscious decision is necessary with regard to working liaisons. Is the planner going to interface directly with the Operations department, or is that relationship to be a function of the maintenance management level to which planning reports? Figure 5-12 depicts the two options:

1. Given indirect liaison, the planner supports only the maintenance manager. He is, in effect, a staff assistant to the maintenance manager. All other liaison is accomplished within the level (both within maintenance and with operations).
2. Given direct liaison, the planner is supporting maintenance management and supervision as well as the operating unit to a maximum degree. This is in keeping with the current team concepts and individual participation and involvement precepts.

Direct liaison is the preferred mode!

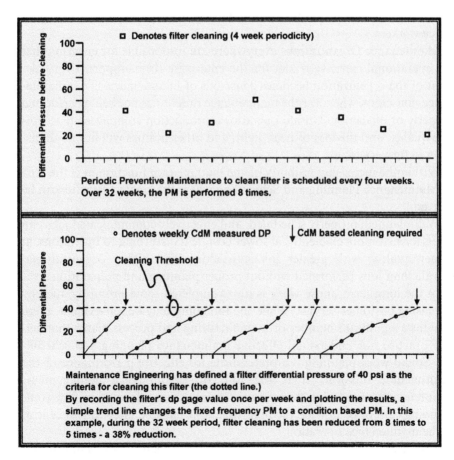

Figure 5-12 Indirect and Direct Liaison

5.1.3.2 Planning and Scheduling: Defining the Role

It seems that everyone today is looking for some ways to improve maintenance. As maintenance assumes an increasing share of manufacturing conversion and upgrade costs, everyone involved wants to ensure that every dollar spent on maintenance is worthwhile. The search for maintenance improvements is centered on finding a program, concept, or approach that will improve the productivity of maintenance labor while at the same time improving production equipment reliability, availability, and productivity. The search must be successful if the manufacturing sector of the economy is to remain viable and competitive.

The "pot-of-gold" can be elusive but finding it begins with sound work planning. Without proper planning and scheduling, maintenance, at best is haphazard; at worst, it is costly and ineffective. The comprehensiveness and sophistication of maintenance Planning and Scheduling is the primary determinant of the level of output and of the degree of productivity of maintenance workers.

Maintenance Organizations everywhere are responsible for ensuring that the operational capacity is met for the enterprise they support. Every element of the organization becomes conscious of maintenance, if not maintenance-conscious, whenever the maintenance function is not being performed properly or efficiently. Unsafe operations, production stoppages, unacceptable quality, and the loss of heat, light, and other utilities will not go undiscovered for very long.

Within the increasing visibility of the maintenance function, it is the area of Maintenance Planning and Scheduling where the greatest benefits can be derived.

Well-planned, properly scheduled, and effectively communicated jobs can be performed more efficiently, at lower cost, less disturbing to operations, at higher quality, with greater job satisfaction and higher organizational morale than jobs performed without proper planning and preparation.

In the long term, more work is also completed more promptly, thereby increasing customer service. While the foregoing may be self-evident, there remains a significant number of manufacturing and process plants that neither practice nor understand effective maintenance planning and scheduling. Sound work planning is a prerequisite for effective performance of the maintenance mission. It is the hub of maintenance management. Maintenance information flows in and maintenance information flows out. Planning cannot be outside the mainstream, it must be integral and central to the maintenance operation.

The primary objective of work planning is to identify all technical and administrative requirements for a work activity and to provide the materials, tools, and support activities needed to perform the work. These items should be provided to tradespersons in an easy-to-use, complete work package.

Effective planning should help ensure that consistent, quality maintenance activities are conducted safely, correctly, and within the allotted time duration. When coupled with an effective scheduling and coordination methodology, delays in performing plant maintenance should be eliminated. Work planning is an evolutionary process that should be periodically assessed through field observation of work being performed and direct feedback from the tradesperson to the planners. An effective planning program should contain the following key elements:

- management commitment, overview, and support to ensure success of the program;
- management direction to ensure appropriate level of detailed work instructions is developed and provided;
- consistency in planning between disciplines to avoid confusion and frustration of work groups;
- thorough reviews by experienced individuals of products produced by the planning group to minimize and eliminate errors;
- feedback from tradespersons and supervisors to facilitate future planning activities;
- use of job history for establishing standard job durations, parts, and consumables for repetitive jobs.

5.2 TPM–RCM–LEAN ORGANIZATIONAL CONSIDERATIONS AND CHOICES

As stated earlier, the combination maintenance and production "team" style of organization is the preferred style for effective TPM execution. How does the introduction of selected RCM techniques affect this arrangement? RCM was originally developed for the aircraft industry where "basic equipment conditions" (no looseness, contamination, or lubrication problems) are mandatory, and where operators' (pilots) skill level, behavior, and training are of a high standard. Unfortunately, in most manufacturing operations that have not implemented TPM, these "basic equipment conditions" and operator skill and behavior levels do not exist. These conditions would almost certainly undermine the success of any RCM application. The introduction of RCM into a TPM environment as a Lean Maintenance element, however, is accorded a greater chance for success because in TPM (a) "basic equipment conditions" are established, (b) equipment-competent operators are developed, (c) combined team members exhibit a posture of stewardship over systems/equipment, and (d) Maintenance Engineering is established for acceptance of "Reliability" Engineering role.

Other considerations for organization of the Maintenance Operation and the Planning and Scheduling function revolve primarily around plant size and combined maintenance/operations team assignment criteria. In the smallest manufacturing operations, there may only be a single person assigned to the planning and scheduling arm of the Maintenance Organization. That person would therefore be responsible for all planning operations, including material identification and staging, work instructions, safety and environmental precautions, and other planning functions. In addition, he or she would be responsible for identifying labor resources, liaison/coordination between maintenance and operations, scheduling and coordination meetings, and work order management, including the maintenance of equipment history. As you can see, these are a lot of responsibilities. Some individual functions of the overall maintenance planning and scheduling process may actually be performed by another individual (e.g., material ID and staging, storeroom personnel; work order data recording, IT or office administrative personnel, etc.) but the lone Planner/Scheduler must retain oversight and control of any functions performed by other personnel.

In small, but not the smallest as just cited, manufacturing plants, a horizontal arrangement is the norm. The horizontal, across facility, organizational structure normally utilizes one Planner/Scheduler performing planning, scheduling, material coordination, and operating liaison for all maintenance work associated with one or more production areas, or for one or more central trade groups. An alternative in large organizations is vertical segregation of planners, schedulers, and material coordinators. The selection of structure is often predicated upon available skills. Planning requires more trade knowledge than scheduling. The latter (vertical) structure can conserve planning capability. Similarly, when the material management system is complex and system knowledge is limited, the material coordinator position should be considered. In either case, it should be centralized to preserve independence. The basic horizontal and vertical organization structures are illustrated in Figure 5-13.

Additional choices in structuring the Maintenance Operation in general, and the Planning and Scheduling function in particular, are all predicated on the complexity of the manufacturing operation and its size (magnitude and diversity of operating systems/equipment). Consolidation of specific functions may be desired in order to maintain adequate control as well as to perform the function effectively. An example would be a Material Control Supervisor, who provides direct oversight of individual material coordinators within each of the Planning/Scheduling teams.

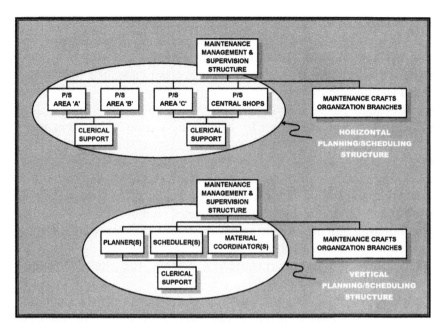

Figure 5-13 Horizontal and Vertical P/S Organization Structures

5.2.1 Where Does the Planner Fit?

It was pointed out earlier that the Maintenance Planner should be on par with the first level maintenance supervisor for whom he or she is planning. Depending again on the size and structure of the manufacturing operation and the maintenance organization, it is strongly recommended that the Maintenance Planner report to the Maintenance Manager or a Maintenance Control Supervisor. In part, this is due to the pivotal role of the planner relative to equipment reliability assurance and in part to the required attributes and knowledge level of the planner. At the same time, it is necessary to consider the roles of others in order to establish where planner responsibility ends and responsibilities of others begin. To be effective, Maintenance Planners need

- recognition that they are planners, not maintenance supervisors; to be a recognized part of the maintenance team; commitment from maintenance and operation management to hold structured backlog review sessions to establish priorities for daily, weekly, down day, and major outage work;
- sufficient lead-time on jobs to plan properly and for getting adequate labor resources;

- to be continuously kept advised on maintenance resource levels;
- to have their relationships with maintenance superintendents and supervisors and with operations clearly defined;
- to have work-requests written by requesters with adequate information and identification;
- an adequate place to work in;
- adequate computer support for developing a comprehensive planning database;
- adequate purchasing support so the planner need only identify required purchases and prepare the necessary purchase requisition, and need not do sourcing, purchasing, purchase order preparation, and delivery expediting;
- adequate storeroom support so the planner need only identify required withdrawals and prepare the necessary stores requisitions, and need not do stock picking, job kitting, and order staging;
- adequate receiving support so that the planner is reliably alerted when purchased items are received;
- adequate reliability engineering support so the planner does not have to develop standard operating and safety procedures, and does not have to devote time on engineering attention to recurring maintenance problems;
- cooperation from maintenance supervisors, tradespersons, and operating supervisors in the effective use and application of efforts put into a meaningful planning package;
- feedback from maintenance supervisors and operating supervisors regarding specific shortfalls in planning packages in order to facilitate future improvements;
- feedback from maintenance supervisors regarding compliance with and exceptions to the weekly plan.

The next and obvious question is, due to the planner's pivotal role, degree of responsibility and high level of maintenance knowledge required, just who is the appropriate person to assign to the planner function?

Under all circumstances, whenever maintenance is performed, it is planned. However, the important questions are: who is doing the planning, when they are doing it, to what level of detail, and how well. Frequently, the wrong person does little planning, at the wrong time, and usually under the pressure of a reactive maintenance environment.

Tradespersons, supervisors, and planners all contribute to the planning process. However, the assignment of para-professionals to the bulk of the planning process is the preferred solution. Job preparation should be a staff function, thereby freeing supervisors and tradespersons, which are line functions, to concentrate their efforts upon the separate duty of job execution.

Planning and execution require different skills. A combination of these skills in one person is the exception rather than the rule.

Separation of planning from execution has long been a rule of good organizational structure. Manufacturing supervision is usually supported by strong staff support groups providing the specialized functions of production planning, scheduling and control; methods, procedures and standards; product quality inspection; operator training; cost control; and inventory management.

5.2.1.1 Erroneous Thinking

Unfortunately, with the arrival of "world-class" philosophies, support functions, and their associated staff (such as maintenance planning) are being drastically reduced in exchange for increased worker involvement, which often includes the absorption of traditional staff duties into the job routines of "work teams." This provides a new argument for those managers with poor understanding of, and appreciation for, the resources required to provide effective maintenance support of operational and production plans and objectives. This approach deprives the Maintenance Operation of an adequate planning function on the grounds that creation or continuation of the function runs counter to modern principles of human behavior. This is based on thinking that, when formal planning (as a separate and distinct function) is eliminated, the task of job preparation and decision making is forced down to the lowest organizational level possible (i.e., the tradesperson). Because this appears to be consistent with TPM and Lean Maintenance concepts of increased worker involvement and self-direction, it is thought that the tradesperson who plans his or her own work is more motivated and therefore more productive.

5.2.1.2 The Reality

The real objectives of increased worker involvement and self-direction postulated by TPM and Lean Maintenance theory are meant to be focused within the line functions of equipment operation and maintenance. When technicians are engaged in planning, they are not executing productive work in the "Lean" fashion and they are not adding value to the production process. Furthermore, people preparing for their own work efforts are less successful than the para-professional in the efficient organization of materials, parts, tools, technical documentation, support personnel (internal and external), and transportation required to execute maintenance work.

In the high-technology processes of today, good maintenance technicians are a precious resource. They are the primary source of effort aimed at preserving reliable productive capacity from industrial equipment and processes. In most facilities, only technicians are contractually authorized to use tools. Yet, numerous studies show that maintenance technicians spend, on average, only two or three hours per day actually applying the tools of their trades. It is imperative, therefore, that traditionally staff-performed activities that consume a technicians time and are not related to equipment reliability or production goals be accomplished by alternative means and alternative parties (support staff).

The aim of effective maintenance planning and scheduling is to optimize the utilization of maintenance resources, contributing directly to equipment reliability. Even when organizations are structured for increased worker involvement and stewardship of their equipment, someone must still be responsible for maintenance planning. The structured planning provided by para-professionals in a staff role within the maintenance structure remains an essential requirement, especially in those organizations that have adopted self-directed/participative team concepts. Some person or group must provide planning support in a methodical manner. Without planning, the resources used in the manufacturing process or in providing maintenance services are being consumed ineffectively. Unfortunately, many managers are slow to acknowledge the need. We return to the original question of just who is the appropriate person to assign to the planner function. The options are as follows.

5.2.1.3 The Assigned Tradesperson

If planning is left up to the employee, it is rarely performed well. It results in delay, wasted effort, and inefficiency. The worker engages in activities that reduce time available for direct work. Indirect activity and travel tend to be very high in proportion to direct work time. Because of his or her position in the organizational hierarchy, the tradesperson is not well postured for many of the liaisons associated with the planning and scheduling role.

5.2.1.4 The Responsible Supervisor (or Team Leader)?

Supervisory activity should be focused on instructions and control of methods, pace, and quality of work. First-line supervisors are organizationally postured to concentrate on immediate problems and have little time to focus on future activities. Given the demands of daily maintenance execution, it is difficult for supervisors to provide both the necessary supervision and coordination of today's jobs as well as the effective preparation of future jobs (tomorrow's and next week's work). One or the other is neg-lected. Given the choice and faced with the daily demands encountered, first-line supervisors concentrate on today's problems. Therefore, it is plan-ning for tomorrow's activities that suffer.

Any time that a supervisor does spend on planning reduces the time devoted to field supervision, thereby defaulting on the prime supervisory responsibility of field supervision; in effect delegating the supervisor's own responsibility to the crew and losing the benefits, which his or her presence and availability bring to job execution.

Supervisors typically plan just prior to job start, thus leaving little time to consider methods, and identify and acquire required materials, tools, sup-port trades, etc. The result is missing materials, delays, incomplete jobs, inef-ficient methods, questions and over staffing.

Nevertheless, because of a lack of maintenance understanding and reluc-tance to increase overhead, first line maintenance supervisors are, in many cases, expected to assume responsibility for job preparation as well as job execution. However, the assumed overhead savings are consumed many times over by resultant losses in job quality, equipment breakdown, and crew productivity.

When planning is handled by a supportive staff function, the first-line supervisor has more time for leading and directing his or her team:

1. More supervisory follow through promotes better quality or work and reduces unnecessary delay and idleness.
2. More time can be devoted to individualized training and instructing of team members, thereby, developing apprentices and lower skilled per-sonnel into "tradespersons" of the highest capability, capable of main-taining today's high technology.
3. More time can be devoted to the practice of good employee relations with reduction of grievances and minimal organizational damage through prompt handling of those grievances that do occur. The resultant is high morale and work spirit of the team further con-tributes to increased team output.

The answer to the original question of just who is the appropriate person to assign to the planner function is: A Well-Trained Planner/Scheduler!!!

Separate the functions of planning and supervision. Planning and exe-cution require different skills. A combination of these skills in one person is the exception rather than the rule. Therefore, planning should be a para-professional, staff function, separate from work execution, although prior experience in work execution is an essential requirement. There are definite advantages that result from this separation.

5.2.1.5 Fostering a Sense of Accomplishment

In planning for others, it becomes most important to clarify objectives. Plans need to be designed which provide satisfactory signals when those objectives are being reached. Means for providing feedback of success or failure must be built into plans for others. Performance of individual members working as a group improves the most when they receive constructive information about their individual efforts as well as the group's success as a whole, particularly if the problems are difficult. Equally useful is personal feedback of one member or another in improving the problem-solving efficiency of all.

Promoting Confirmatory Behavior: The confidence of planners in the adequacy of their plans can be communicated, while reservations of the doer need to be brought into the open and discussed with the planners. If the doers believe that the plans are unrealistic, even is the plans are actually sound, the doers may behave in a way to confirm their own beliefs. Planners need to share with the doers the reasons for their optimistic expectations.

Promoting Commitment: Doers can be consulted at various stages in the planning process. Wherever possible, the ideas of the doers can be incorporated in the plans, or when such ideas of the doers can be incorporated in the plans, or when such ideas are unusable, the reasons can be discussed with doers. The plans can provide for some discretion on the part of the doers to modify noncritical elements of the plans, thus increasing the feeling that the doers have some control over the fate of the plans, and consequently, responsibility for the successful execution of them. Built into the planning process can be provisions for periodic feedback of evaluation from the doers.

Providing Flexibility: While those who plan for themselves display more initiative in making needed changes in a plan, specific provisions to encourage such initiative are required when planning for others. It can take the form of an instruction asking doers to regularly report back to the planners whether the plan is working satisfactorily and where it may need modification. Planners may need to allow for a completely unforeseen contingency, the critical event whose occurrence could not have been anticipated, expected by the general instruction to the doers to modify the plan in a suitable way, if such and unsuspected event should occur.

Promoting Understanding: Understanding of the plan can be fostered by ensuring that the plan itself has been created in a way to minimize its ambiguities. Repeating instructions may increase reliability and understanding. Crosschecks and tests of the plan's clarity before it is presented to the doers may be helpful also to ensure that the plan's instructions are simple enough to be understood by the least capable doer. If he or she can understand the stages of the plan, all others can also understand it.

Maximizing Effective Use of Available Labor Resources: This is a matter of making the most of what is available by whatever means the situation allows. Latitude should be commensurate with capacity. Planners need to incorporate allowances for individual differences in their plans for others.

*Improving Communications:*The doers must have (and feel that they have) ample opportunity to question the planners for clarification of the presentation of the plans. The planners need to attend to whether or not they are overestimating or underestimating the ability of those who are receiving the plans to comprehend them. The planners need to judge whether they are transmitting too much or too little too fast or too slowly, with too much or too little enthusiasm, with too much or too little confidence.

Minimizing Competition: The planners need to avoid creating a situation in which the doers see themselves in a zero-sum game with planners. If the plans succeed, the doers must not lose status, prestige, power, or material benefits. Conditions must be established in which the doers share with the planners the same super ordinate goals. The division of labor should be seen as benefiting the doers as well as the planners.

Planning performed by a separate staff group has the added advantage of an overall functional perspective. Priorities, labor resource loading, and management reports are better coordinated through a controlled planning function. Several jobs can be planned more efficiently by a focused function than by an assigned tradesperson planning one job at a time. Therefore, the answer is that they all have a role to play Tradespersons, Supervisors and Planners, all contribute to the planning process.

Regardless of the organization used there should always be a current and complete organizational chart that clearly defines all maintenance department reporting and control relationships, and any defined/designated relationships to other departments. The lines of communications between staff and line functions as well as between departments, although less formal, should always be open. The Maintenance Organization should clearly show responsibility for the three basic maintenance responses: routine, emergency and backlog relief. The role of planning and scheduling is critical for the effective utilization of the maintenance line-resources.

6

Performing the Planning Function

As stated earlier, whenever maintenance is performed, regardless of organization style, it is planned. However, the primary questions remain: who is doing the planning, when are they doing it, to what degree, and how well? Because planning and scheduling functions are at the hub of all maintenance activity coordination; these functions rate a professional grade classification and assignment of only professionally trained, experienced and qualified personnel.

- Maintenance Planning is the advance preparation of selected jobs so that they can be executed in an efficient and effective manner when the job is performed at some future date.
- Maintenance Planning is a process of detailed analysis to first determine and then to describe the work to be performed, by task sequence and methodology.
- Maintenance Planning provides for the identification of all required resources, including skills, crew size, labor-hours, spare parts and materials, special tools and equipment.
- Maintenance Planning includes developing an estimate of total cost and encompasses essential preparatory, postmaintenance and restart efforts of both operations and maintenance.

The Maintenance Planner must have the requisite personal skills as well as professional skills derived from experience and thorough, comprehensive training in order to execute "professional" Maintenance Planning. When these attributes are in place, the effective utilization of maintenance personnel can be increased by as much as 65% and job execution time can be reduced by as much as 40 to 50% (see Figure 6-1).

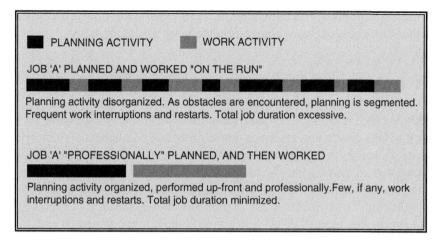

Figure 6-1 Professional Planning Saves Time (and Money!)

6.1 PRINCIPLES OF PLANNING

Perhaps more properly the listing that follows might be referred to as principles for the Planner to strive for in his or her day-to-day planning activities:

- understand the department's mission in relation to the objectives of the company;
- always be aware of the magnitude and trend of backlog;
- quantify the magnitude of the resources effectively available to apply toward relief of the backlog;
- establish a plan for the allocation of available resources to a balanced workweek, considering both long-range importance and short-range necessity;
- categorize work consistent with planned resource allocation categories;
- assign a planning priority (within job priority and category) to each job;
- break each job into logically sequenced tasks/activities;
- prepare a "Planning Week" schedule by phases of work planning and by task to determine progress toward completion of each week's work planning;
- work to meet this schedule. Protect it. Do not superimpose new work unless that new work represents an overriding course of action for work planning (long and short range);

- measure progress and contribution. Don't spin your wheels on efforts that don't move toward completing the week's work planning.

6.1.1 Managing the Backlog

Maintenance is managed by managing the backlog. It is impossible for a facility to be proactive if resources are not kept in balance with the workload. If the overwhelming majority of jobs are not planned, there is no effective way to know what the magnitude of the backlog is and therefore it cannot be managed. Backlog is also an indicator of resource augmentation—through overtime, staff expansion or outsourcing. Backlog Management begins with the Maintenance Planner and culminates with the Maintenance Scheduler.

Just what is backlog? In the simplest of terms, backlog is work that has not yet been completed. When a work order is generated, it contains, among other information, a priority code. The work order, or job, priority determines when the work will be planned, scheduled and performed relative to other jobs. When the planner has completed all of the planning functions for a work order, the job is ready to schedule or *available*. If the work planning is not completed, the work order is not ready to be scheduled or *unavailable*. Together these two classifications of work order status constitute the backlog. Backlog is measured in time, normally hours that are translated into labor weeks. The hours of backlog are the resource, or labor hours that the work is estimated to take. The normal range for backlog, based on 80 to 90% of jobs being planned, is considered to be:

Ready available backlog is equal to 2–4 weeks of labor hours
Total backlog (available and unavailable) is equal to 4–6 weeks of labor hours

The reason that the norm is defined as being based on 80 to 90% of jobs being planned is because, for unplanned jobs, the estimated labor hours are not yet known. For example, 10% of the backlog (number of work orders) is unplanned and the planned portion (90% of the work orders) of the backlog is 1800 hours. The assumption is made that the unplanned portion consists of work similar in labor hour requirements to the 90% of the work that is planned, in this case, unplanned is equal to 200 hours, for a total backlog of 2000 hours. The calculation is:

1800 hours divided by 90% equals 2000 hours (100% or Total Backlog), and 2000 minus 1800 equals 200 hours (total minus planned equals unplanned)

Obviously, if the planned work is very much below 80% of the total, confidence in the estimation of the unplanned work becomes lower and as a result, control of the backlog is lost because the size of the backlog is unknown.

When equating the hours of backlog to weeks of backlog, straight time and scheduled overtime are considered. For example if the average maintenance work force available during any given week is five tradespersons with 2 hours per person scheduled for O/T, then one week of backlog is equivalent to five times 40 (hours) or 200 hours plus 10 hours O/T = 210 hours/week available capacity. Backlog is referred to in weeks simply because it is more understandable than four or five figure hourly references.

The Maintenance Planner must prepare a weekly Backlog Status Report. The report can contain as much or as little information as desired, but as a *minimum* must include the following:

- total backlog (total labor hours of all open work orders in the CMMS);
- ready backlog (total labor hours of work orders ready to schedule);
- age of backlog;
  - < 1 month
  - 1–2 months
  - 3–6 months
  - > 6 months
- trend charts
  - total labor hours completed versus # of backlog work orders completed
  - total available labor hours
  - total scheduled overtime hours

CMMS can be used to automatically generate and disseminate the (planner reviewed and approved) Weekly Backlog Status Report. The Planner should provide a copy of the report to all attendees at the Weekly Maintenance Schedule Planning Meeting prior to the day of the meeting.

In addition to being available or unavailable, backlog also has the characteristic of age. Age simply refers to how old the work order is. The primary determinant of ageing is the work order priority. Obviously the higher priority work orders will be planned before the lower priority ones, therefore one would expect the "youngest" work orders in the available backlog to be predominantly made up of high-priority work orders. As a management tool, work order age is important in decision making regarding the cancellation or re-prioritization (to a higher priority) of older work orders in the backlog.

Appendix D, Section D1, provides a short tutorial on the use of control charts as an aid to backlog management. Exhibit C-8 of Appendix C also contains a Backlog Worksheet that can be utilized in backlog management.

In order to manage and control backlog to predetermined levels, i.e., 2 to 4 weeks of ready available backlog, the Maintenance Planner and the Maintenance Scheduler (if planning and scheduling are assigned to separate people) must work together closely. They must know the level of resources (personnel) available each week. If there is a decrease in available resources (vacation, training, etc.?) there will be corresponding increase in backlog. Without prior knowledge of resource levels, the backlog could go out of the control band. Backlog trends up or down should be investigated if the cause is not known in advance. It is important to know whether it is a temporary trend or a permanent one (e.g., a continuing trend of decreasing backlog is very likely an indication that the size of the maintenance group is too large for the maintenance workload). For the planner and the scheduler to manage backlog based on available resources, they must be apprised of resource levels on a weekly basis. Each maintenance supervisor must be required to submit weekly manning level reports to the maintenance scheduler in order for the scheduler to know how much of the planner's available work can be scheduled. Appendix C, Exhibit C-2 illustrates a "Weekly Resource Report" reporting form that provides all of the information necessary for this weekly reporting requirement.

6.1.2 Criticality and Prioritization

The Maintenance Planner initiates job planning based on work orders received and the coded information on the work order. The coded information includes work type, work category, work classification and perhaps others, but initially the primary interest of the planner is focused on the *priority* that has been indicated for the work (*priority: something given or meriting attention before competing alternatives*).[3] Job priority is the determinant for sequencing work planning: the highest priority jobs are the first to be planned and the first to be scheduled. Job priority is assigned by the originator of the work order (request). The priority is based on equipment criticality and on the type of work to be performed.

The first of these, Equipment Criticality, ranks each piece of equipment in relation to its impact on the production process. Assignment of equipment criticality should be made jointly by Operations (Production) and Reliability Engineering. If equipment criticalities have not been assigned, it is the responsibility of the Maintenance Planner to ensure that they get assigned, a responsibility where the planner's tact, perseverance and persuasiveness can be put to the test. The generation of an equipment listing with

---

[3] *Merriam-Webster's 11th Collegiate Dictionary.*

"tentative" criticality codes (see Table 6-1) assigned by the Maintenance Planner might expedite the process.

The second, Work Type, is determined by (1) work class and (2) work category. Work class is a dynamic attribute such as breakdown repair or repair of a potential failure (failure is imminent; e.g., an overheating bearing) while work category is a fixed attribute such as preventive maintenance or equipment alteration. The work category is also defined, in part, by the purpose of the work such as correction of a safety problem or improvement of equipment efficiency (economics).

A ranking index for maintenance expenditures (RIME) has been developed to help maintenance departments do a more equitable and logical job of controlling maintenance expenditures. The RIME index consists of: (a) equipment criticality, relating to equipment capacity and reliability, estimated repair costs and impact on the process, and (b) work class, which takes into consideration safety hazards, operating costs, and labor.

Combining these two RIME elements provides a better determination of which maintenance jobs should be scheduled for completion first. A comparison of job RIME numbers will indicate which jobs are essential and which ones can wait. Application of the RIME index results in better maintenance decisions and leads to better planning. The equipment criticality

Table 6-1
Equipment Criticality (Fixed)

Criticality	Description
10	Shuts down entire plant—e.g., utilities
9	Shuts down more than one line—key production equipment
8	Not spared production equipment/shuts down one line
7	Mobile equipment—e.g., fork truck, transporters
6	Spared production equipment/not spared support equipment—product can be made on one or more lines
5	Support equipment spared
4	Infrequently used production equipment
3	Miscellaneous equipment—e.g., water cooler, windows, cafeteria
2	Roads and grounds
1	Buildings and offices

codes are the same as previously defined in Table 6-1. The work class descriptions are provided in Table 6-2. Work Class ranks each job in relation to each maintenance job or project. Following Table 6-2 is a series of tables in Table 6-3 (a through d) with "suggested" priority assignments for the RIME matrix. It is emphasized that these are suggested priorities. The relationship between equipment criticality and work class as they relate to work order priority assignment is a local policy issue for individual plants. Whatever the priority correlation with the RIME Index, it should be spelled out in the plant's Work Management Standard Operating Procedure (SOP). In addition to this "local policy" characteristic, the RIME Index assigned priorities must also have built-in flexibility sufficient to allow any values to be overridden by the Plant Manager, the Maintenance Manager and/or the Production Manager (Table 6-2).

Index #	Work Class Definitions
10	*Breakdown—Real Safety Regulatory Compliance*: Equipment stoppage during plant operation. No production output. Immediate threat to *life or limb*. Environmental impact or a serious citation that may shut down a piece of equipment (e.g., high-voltage panel not protected).
9	*Product/Quality Loss*: A malfunction that does not result in line shutdown but causes intolerable product/quality problems (e.g., code date, open flaps).

Table 6-2
Work Class (Dynamic)

Index	Description
10	Breakdown/Real Safety/Regulatory Compliance/Quality
9	Product/quality loss
8	Potential breakdown
7	Preventive maintenance
6	Working Conditions/Safety/Security
5	Shutdown work
4	Normal maintenance
3	Projects and experimental
2	Cost reduction
1	Spare equipment/parts

Table 6-3
RIME Index Priority Assignments

(a)

Work/Job Priority Validation

Class	Criticality									
	10	9	8	7	6	5	4	3	2	1
10	E	E	E	1	2	2	2	2	2	2
9	E	E	1	1	2	3	3	N/A	N/A	N/A
8	1	1	1	2	3	3	3	3	3	3
7	4	4	4	4	4	4	4	4	4	4
6	4	4	4	4	4	4	4	4	4	
5	6	6	6	6	6	6	6	6	6	6
4	4	4	4	4	5	5	5	5	5	5
3	5	5	5	5	5	5	5	N/A	N/A	N/A
2	5	5	5	5	5	5	5	5	5	
1	N/A	N/A	N/A	7	7	7	7	7	7	7

(c)

Criticality (horizontal values for Table (a))

Criticality	Description
10	Shuts down entire plant—e.g., utilities
9	Shuts down more than one line—key production equipment
8	Not spared production equipment/shuts down one line
7	Mobile equipment—e.g., fork truck, transporters
6	Spared production equipment/not spared support equipment—product can be made on one or more lines
5	Support equipment spared
4	Infrequently used production equipment
3	Miscellaneous equipment—e.g., water cooler, windows, cafeteria
2	Roads and grounds
1	Buildings and offices

(b)

Work/Job Priority

(cell inputs for Table (a))

Priority	Title
E	Emergency
1	Urgent
2	Critical
3	Rush
4	Essential – but deferrable
5	Desirable
6	Shutdown
7	Routine

(d)

Class of Work (vertical values for Table (a))

Class	Description
10	Breakdown/Real Safety/Regulatory Compliance/Quality
9	Product/quality loss
8	Potential breakdown
7	Preventive maintenance
6	Working Conditions/Safety/Security
5	Shutdown work
4	Normal maintenance
3	Projects and experimental
2	Cost reduction
1	Spare equipment/parts

8   *Potential Breakdown*: an identified problem which must be corrected as soon as production is curtailed (e.g., conveyor belt splice is tearing apart).

7   *Preventive Maintenance*: Repairs that are identified and performed in a preplanned mode to avoid breakdown and all normal work orders generated from performing PM work orders and inspections (e.g., inspections and running adjustment).

6   *Shutdown Work*: Work which is not critical enough to require immediate shutdown but must be performed only during a planned shutdown period (e.g., major overhaul jobs that are large in scope).

5   *Normal Maintenance*: Routine work that can be planned, scheduled and completed without disrupting planned production output (e.g., rebuild gear case).

4   *Working Conditions—Safety and Security*: Any change in physical environment that is either aesthetically pleasing or motivational, minor safety and security work (e.g., repaint, repair and office door/lock).

3   *Projects and Experimental*: Engineering projects which are requested to modify design, or improve reliability of equipment or processes (e.g., installation of hot air sealer).

2   *Cost Reduction–Corrective Maintenance*: Work that results in operational changes that will reduce unit costs and does not fall into one of the higher classifications. Replacement of a defective component to eliminate or reduce repetitive repairs (e.g., metalizing a cylinder rod, replacing an open bearing with a sealed bearing).

1   *Spare Equipment/Parts*: Fabrication of multiple spare repair parts (e.g., turning multiple spare pump shafts in a lathe).

The RIME index can also be useful to the maintenance planner as an evaluation tool in determining the validity of originator assigned job priorities, although, most often, the job priority is self-evident and does not require validation by the planner. However, if there is a question regarding job priority assignment, the planner must *always* direct any questions back to the work order approving authority for resolution.

This may all seem quite convoluted: work type, work class, work category, work purpose and seemingly on and on. The graphic in Figure 6-2 may help to make the relationships and usage of work/job attributes a little clearer.

Job Priorities are assigned by the WO approval authority. Priorities rank each job in relation to others and establish a time frame within which jobs should progress. Priority code systems can range from 3 or 4 rankings to as many as 15; too few rankings do not provide adequate separation to aid in decision-making, while too many can become cumbersome and difficult to manage. The objectives of assigning work priorities are to:

- separate plannable from unplannable work;
- facilitate sequencing of work order planning and execution;
- provide indication of lead time available to plan.

Table 6-4 illustrates a recommended Priority Coding system. Following Table 6-4 are complete condition descriptions of each indexed priority code:

E   *Emergency*:   *Must be performed immediately.* Higher priority than scheduled work, critical machinery down or in danger of going down until requested work complete. "E" to be used only if production loss, delivery performance, personnel safety (new and imminent), equipment damage or material loss are involved and no bypass is available *Start immediately and work expeditiously and continuously to completion, including the use of overtime without specific further approval.* Only personnel authorized to approve overtime can assign "E" priority

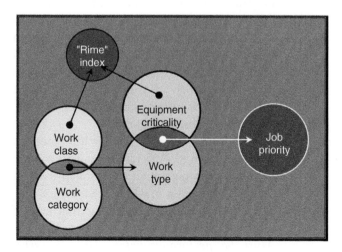

Figure 6-2 Relationship of Priority to Work and Equipment Classifications

Table 6-4
Job Priority Assignments

Index	Title	Job Planning
E	Emergency	Un-plannable*
1	Urgent	Un-plannable*
2	Critical	Usually un-plannable*
3	Rush	At least partially plannable
4	Essential—but deferrable	Plannable
5	Desirable	Plannable
6	Shutdown	Plannable
7	Routine	Plannable

*Un-plannable from the standpoint of time. If the work has been performed previously and planning completed, all work planning information should be made available to the work crew.

to work orders. Emergency work order reports will be sent to plant manager for review.

1   Urgent   *Needed within a few hours, by end of shift at latest.* It is the opinion of the work order approval authority that the work must be completed as soon as possible but does not constitute an emergency requiring immediate attention. Maintenance personnel are assigned as soon as available without halting a job already in progress. Overtime approval is not implied by "1" priority. Overtime authorization must be obtained by special request, given specific circumstances, from the authorizer or designated authority. Priority "1" should be used for equipment that is down or in danger of going down and which affects ability to produce desired product mix or renders plant void of backup capacity in event of subsequent failure.

2   Critical   *Needed within 24 hours.* Similar to urgent (1) jobs but with less urgency. Typically good work to leave for off-shift coverage personnel. Controlled use of "E," "1" and "2." Priorities must be reserved for truly critical situations or they diminish planning and scheduling effectiveness.

3   Rush   *Must be performed before end of current week.* Normally, this work will be scheduled to start within

48 hours after receipt of the work order. Priority 3 jobs (as well as "E," "1" and "2" jobs) cannot be effectively planned before scheduling. All jobs should be assigned priority "4" or higher whenever possible. Priority 3 jobs will be used as fill-in work for personnel responding to emergency and urgent work orders or will be forced into the current week's schedule, "bumping" a properly planned job already on the schedule. Performance on the job will be measurably less efficient and cost will be measurably greater than if planned (Priority "4" or "5").

4	*Essential but deferrable*	*Must be performed before the end of next week.* This work can be effectively planned. It will be scheduled next week (as opposed to being scheduled in the order of request date). Realize that all requests cannot be completed next week. Priority "4" requests delay the completion of previously requested work of lower priority and drive up the cost of requested work, as overtime will be requested to meet the time/demand constraints implied by priority. Use Priority "5" if possible.
5	*Desirable*	*Designates desirable but deferrable jobs. Can be completed anytime within the next few weeks,* and can therefore be scheduled on the basis of first requested-first scheduled. A desired completion date may be indicated by the originator. Weeks of backlog report provide the current wait to be anticipated on Priority "5" jobs. Overtime will not be used on Priority "5" jobs unless the work must be performed on a nonoperating day or aging of the request exceeds six weeks.
6	*Shutdown*	*Work requiring programmed shutdown.* Work orders in this category are accumulated for shutdown planning.
7	*Routine*	*Used exclusively for routine work.* Usually assigned to standing work orders, "7" is not associated with normal day-to-day work order requests.

Assuming that priority codes E, 1 and 2 (Emergency, Urgent and Critical) are well controlled and used only for truly critical situations, the planner is predominantly involved with work orders of priorities 3, 4 and 5. Priority 6 (shutdown work orders) is discussed later on in Chapter 8 and priority 7 (routine work orders) is dealt with differently in the Lean Maintenance Environment and will be discussed later in this chapter. Within priorities 3, 4 and 5 are requested completion times of 3 to 4 days, 5 days to 2 weeks and

2 weeks to 5 or 6 weeks. It is this latter category that comprises the majority of backlog work. The age of priority 5 work backlog is the primary determinant for scheduling sequence. Work Type and Equipment Criticality are secondary scheduling determinants for work orders of the same age.

Special Note on Safety Jobs: Safety jobs are highlighted on the work order by reason code "S." Safety priority is always high. If there is an active hazard requiring immediate attention and correction, the work order should also be given an "E" Priority. All safety jobs not given an "E" priority should be handled on an individual basis and assigned one of the applicable priority codes above.

6.1.3 The Work Order

The heart of Maintenance Planning beats only as strong and healthy as the Maintenance Organization's Work Order System functions. Knowledge and control of the tasks to be performed are essential to all functions. For maintenance, the source of such knowledge and control is the *Work Order*.

6.1.3.1 Work Order Types and Formats

There are three basic work order types or formats:

1. Formal Work Order
2. Standing Work Order (SWO)
   a. Daily Checks/Routines (SWO subtype)
3. Unplanned/unscheduled (break-down) Work Order

The Formal Work Order is used for those jobs that enter the backlog and that can be planned each time they occur. In designing the formal work order, a decision is required: *single-trade format* or *multitrade format*.

- *Single-trade format.* The single-trade format is an internal maintenance department document (usually a copy of the child work order and attached to the parent work order) assigning portions of job to trade other than parent and aids in coordinating multitrade jobs. Jobs are

planned and scheduled in the same manner and carry the same reference number as the parent work order.
- *Multitrade format.* A multitrade work order details work sequence by craft in the planning section of the work order. For control of activity, backlog and performance by craft, it is essential that the computer system treats each sequence line of multitrade work order as separate jobs. Manual systems (as opposed to CMMS) often lose control by trade if a multitrade work order is utilized.

Standing Work Orders (SWOs) are used primarily for periodic jobs performed on a predetermined schedule, in a standard pattern and involving a standard amount of work. It is unnecessary to prepare unique work orders each time these jobs become due. SWOs are commonly established to absorb charges for the following types of work:

1. Routine inspection, lubrication, adjustment and service of equipment.
   a. Includes most Autonomous Maintenance (Operator performed)
2. Repetitive service work:
   a. Re-lamping
   b. Trash removal
   c. Janitorial duties
   d. Material and supply transportation
   e. Fire checks.
3. Bench work with visible backlog
4. Daily checks/routes
5. Minor, short- and low-cost repairs by a technician assigned to a specific unit where no equipment history is required.
6. Nonmaintenance activities:
   a. Safety and union meetings
   b. Training.

SWOs are assigned to individual cost centers, but should also permit separation of expenses by equipment number when issued for work exclusively on a particular machine. Any corrective work should be captured on a formal work order.

Repetitive service work is most economically controlled without the use of a formal WO. For each task in a day's work of this type, the effort and expense involved in formal WO are not justifiable. The cost of execution is not much more than expense of preparing, coordinating and controlling formal WO. SWOs should be numbered separately from the WO sequential numbering scheme and the date performed code appended to control and track their cost and performance. By identifying work of this nature and controlling it with standing work orders, a large quantity of minor tasks can be controlled promptly, effectively and economically. However, only a minimum number of the most useful and necessary standing work orders should be established.

Most, if not all, Preventive Maintenance tasks are automatically generated by CMMS. Preventive Maintenance tasks are pre-planned, thus limiting the planner's involvement. However, any corrections or modifications required for pre-planned Preventive Maintenance tasks should be reported to the Maintenance Planner, who is then responsible for developing any changes required. The Maintenance Scheduler lists those CMMS generated PMs on the Maintenance Weekly Schedule in order to coordinate equipment availability, track performance, schedule compliance and generate equipment history. The Preventive Maintenance Group Supervisor is responsible for ensuring completion of all Preventive Maintenance on schedule. Additionally, the supervisor determines which routine PMs are to be performed by the Production Department Autonomous Maintenance Teams. Through coordination with the maintenance trades assigned to these teams, the PM Group Supervisor assigns them minor and routine PM tasks based on the level of training and demonstrated skills of autonomous maintenance equipment operators. The PM Group Supervisor may assign nonminor and nonroutine tasks to the Autonomous Teams when and if their training levels support such work and the team's resource levels are sufficient to perform the additional maintenance.

SWOs should be reviewed, validated and redefined monthly, or at some other predetermined frequency. Estimated costs are identified and authorized for each by designated authority. Charges are controlled against monthly budgets. Cost variance reports for each standing work order should be closely reviewed. Sharp fluctuations or excess charges should be investigated to determine cause and initiate corrective action. Daily checks/routes that simply employs a one- or two-line entry into a chronological log represents another variation of the SWO. Typically, these are once per shift or once per day readings taken from equipment instrumentation and similar high-frequency, low-effort work. Each sheet (template) is assigned a unique number and the date or period (normally one week) covered by the check/route entries is entered when the sheet is completed. Data from completed sheets should be entered into CMMS equipment history by the Planning and Scheduling group's administrative clerk/aide. Costs are handled in the same manner as SWOs.

Unplanned/unscheduled work orders are used for those jobs that must be completed within the next day or so, without the opportunity to plan completely or plan at all, as they require immediate attention. These work orders will have a high-priority rating. Completion and closeout of this type of work order requires documentation of all actions taken. Labor and material costs are logged back to the equipment and a copy of the details of the failure and repair are forwarded to Reliability Engineering for further failure analysis.

6.1.3.2 Work Order System and Work Flow

The work order should be viewed as a system, not as a form or procedure. The work order provides the means for requesting maintenance service, and thereafter planning, scheduling, controlling, executing and reporting the work performed.

An effective WO System

- screens out the unnecessary and unimportant;
- establishes responsibility;
- reduces mistakes;
- provides an understanding of what is to be done;
- provides a means of charging labor, material and outside services to the equipment owner;
- serves as an authorization document;
- is a source document for maintenance cost and performance control;
- is the drive wheel of integrated maintenance management.

Through proper use of the WO system, accurate work backlogs are established, job preparation is facilitated, control of maintenance work is enhanced, equipment histories are created and optimum effectiveness of maintenance work groups can be achieved. There are two approaches to administering a WO system. The first is characterized by the use of work requests, which are submitted to the maintenance planner who is then responsible for generating, including the coding of all data fields, the work order. The second, which is used in capturing unplanned/unscheduled work, is characterized by initiation of the work order directly by the originator. The completed work order is then submitted to the planner for additional coding and CMMS entry.

While the work order is used in conjunction with other functions, such as time keeping, inventory control and purchase orders, as input to the accounting cost allocation system, the work order is first and foremost a maintenance management tool. Before development of CMMS, maintenance costs were charged directly to cost control accounts (in varying degrees of specificity) via the other three functions listed above. Nonmaintenance costs are still widely accumulated in this manner. Combined usage of the work order system for cost control as well as operational control came about as an enhancement to tie the two systems together and to avoid duplicity of input effort. Many installations still do not merge the two systems.

Work Order Control System

The work order is a control document serving three basic functions:

1. Definition and authorization of work to be performed.
   a. Systematic screening and authorization of work requested in respect to work type, location, urgency and cause.
2. Planning and control of work to be performed.
   a. Record and measure amount of work coming in (inputs)
   b. Assign priorities
   c. Provide information needed to plan, schedule and coordinate methods, materials, labor
   d. Supply supervisors and technicians with job instructions and estimates of the time required
   e. Accumulate information on job progress
   f. Record and measure amount of work completed (outputs)
   g. Control manning levels to balance output with input
3. Maintenance History accumulation and input.
   a. Develop time estimates for repetitive work
   b. Perform Maintenance Optimization and Root Cause Failure Analysis (RCFA)
   c. Cost and performance measurement and improvement
   d. Analyze costs by job, equipment and cost center
   e. Improve planning, scheduling and preventive maintenance

Maintenance Work Order Data

The matrix of data in Figure 6-5B, showing all work order fields, responsibilities and terms for data entry, apply to plannable, comprehensive work orders. Some simplification is provided for standing and unplanned/unscheduled work orders.

The workflow of the work order should be consistent and should not be bypassed for minimal speed gains or convenience. Even unplanned/unscheduled work originated by emergency or urgent priority work order should pass through the maintenance planner when completed. Delivery of high-priority work orders can be made by hand and in person to expedite handling, but to ensure proper logging, handling and follow-up, all work order priorities *must* go through the planner. In addition, it is just possible that the repair called out in an emergency work order has been performed before and archived plans may be available. In the long run, consistent handling may actually speed up the repair process—even of high priority, normally unplanned, work. Figure 6-3 illustrates an effective workflow scheme.

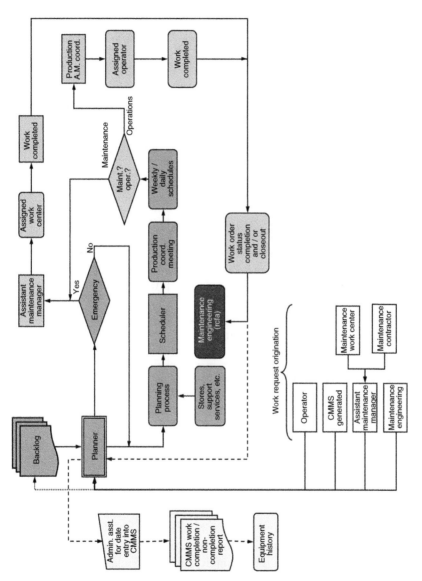

Figure 6-3 Planned and Lean Work Flow

Table 6-5A
Minimum Work Order Information

Basic Work Order Information		
Job Definition	**Planning, Scheduling and Control**	**History-Analysis**
Unique Job Number	Trade Code(s)/Skill Requirement	Actual Labor Costs
Equipment Number	Assigned Organization	Actual Materials Cost
Equipment Description	Job Plan	Actual Work Performed
Location	Labor Sequence	Failure Code(s)
Account Number	Bill of Material	Action Taken (Verb)
Project Number	Labor Cost Estimate	Component Affected (Noun)
Work Type / Safety?	Materials Cost Estimate	Condition Found (Adjective)
Work Purpose	Status Code	
Work Requested and Action taken	Estimated Duration	
Job Priority	Schedule Sequence	
Equipment Criticality	Related Work	
Approval/ Authorization		

6.1.3.3 Coding Work Order Information

It is clear from Tables 6-5A and B that, unless some sort of shorthand for entering the required information on the work order, the work order form will need to have 8 to 10 pages. Fortunately, there are (relatively) universal codes, numerical, alphabetical and standardized phraseology and acronyms, that are used to annotate work orders. Whether you use the coding provided here, or formulate your own, work order codes and definitions *must* be widely published and *must* be used by *everyone* that will encounter the work order. The mandatory and universal use of standardized coding (and

Table 6-5B
Minimum Work Order Data

Field	Responsibility	Timing
To Define the Job:		
Unique work order number	Automatic	Origination
Date of request	Automatic	Origination
Time of request	Automatic	Origination
Requestor	Originator	Origination
Equipment number	Originator	Origination
Equipment description	Automatic	Origination
Equipment location	Automatic	Origination
Work requested	Originator	Origination
Work type	Originator	Origination
Account number	Automatic	Origination
Project number	Originator	Origination
Job priority	Originator	Origination
Current equipment status	Originator	Origination
Time equipment went down	Originator	Origination
Requested completion date	Originator	Origination
Date equipment available	Originator	Origination
Approval	Proprietor	Origination
To Prepare the Job:		
Planner	Planner	Preparation
Skill code	Planner	Preparation
Crew / task sequence	Planner	Preparation
Crew interfaces	Planner	Preparation
Crew size	Planner	Preparation
Estimated man-hours	Planner	Preparation
Assigned organization	Planner	Preparation

Table 6-5B
Continued

Job code	Planner	Preparation
Bill of materials	Planner	Preparation
Job status	Planner	Dynamic
Cost estimate (labor, materials)	Planner	Preparation
Schedule sequence	Planner	Scheduling
Job schedule	Planner	Scheduling
Suggested manpower deployment	Planner	Scheduling
Equipment back on line	Operator	Completion
Date back on line	Automatic	Completion
Time back on line	Automatic	Completion
Actual work performed	Maint. Trade	Completion
Action taken code (verb)	Maint. Trade	Completion
Component code (noun)	Maint. Trade	Completion
Condition code (adjective)	Maint. Trade	Completion
Failure cause code	Maint. Trade	Completion
Job completion	Maint. Trade	Completion
Actual labor (hours and $)	Automatic	Completion
Actual material use and $	Automatic	Completion
Job quality acceptance	Originator	Completion

abbreviations or acronyms) throughout the plant cannot be overemphasized. Even if only one participant deviates from standardized practices, the work order control system is compromised and history will be inaccurate. Following then are some suggested work order coding schemes.

Work Order Status Codes

Status codes are for the tracking of work orders as they progress through the life of the process. Only work orders that have reached the status of ready to be scheduled or "available" are considered for scheduling and execution. Both numeric and alphabetic codes together with their description and person responsible for code assignment are provided in Table 6-6.

Table 6-6
Work Order Status Codes and Descriptions

Alternative Codes			
Numeric	**Alpha**	**Description**	**Responsibility**
01	P	Waiting to be planned	Data Entry
02	E	Waiting for engineering/ design	Planner
03	A	Waiting for management approval	Planner
04	PF	Deferred—Pending Funding	Controller
05	PO	Waiting for PO to be issued	Planner
06	M	Waiting for materials	Planner
07	FP	Received and ready for further planning	Receiving clerk
08	DW	Waiting for weekend downtime	Planner
09	DP	Waiting for programmed downtime	Planner
10	DS	Waiting for "no production scheduled"	Planner
11	DO	Waiting for downtime window opportunity	Planner
12	R	Ready to be scheduled	Planner
13	FI	Ready for fill-in assignment	Planner
14	S	Scheduled	Planner
15	CM	Completed-materials/ rebuilding pending	Planner
16	CP	Completed-pending print revision	Planner
17	C	Closed-moved to history	Planner

Definitions of Work Order Status Codes

Code	Description: Definition
01/P	Waiting to be Planned: This code is recorded as a work request is received. Even though no estimates have been posted, it is imperative that all requests be captured and added to the backlog immediately. The goal is to keep the WO system as near to "real time" as possible. In some cases, a backlog builds in front of the planners. This code provides awareness of that condition. Without it, awareness is sacrificed. Whoever enters the work request into the computer system is responsible for entering the initial status code or it can be set as the default code. All but the emergency and urgent work requests pass through this status. These two priorities do not pass through the planner unless their priority is downgraded. They never reside in backlog, so their status need not be tracked. They are either open or completed.
02/E	Awaiting Engineering/Design: Upon review of a work request, planners may determine that engineering and/or design is required before that job can be planned. Until the required design documents are returned by engineering, the request resides in status code -02.
03/A	Awaiting Management Approval: More costly jobs require higher authorization levels for approval. Planners or engineers develop the necessary costs estimates and forward the necessary documentation to the appropriate individual(s) for approval. With approval by the required authorization level, the work request becomes a work order.
04/PF	Awaiting Funding: Cost estimates may be more than the current budget allows. Until funding is determined, the work order resides in this status.
05/PO	Awaiting Purchase Order Issuance: This code is a substatus of awaiting material. At times, the internal processing time consumes as much lead-time as external processing and delivery time. Status code-05 captures this condition. Management focus may be directed to the capacity of purchasing/buying resources. If processing time is under control, the status code is optional–a local judgment.
06/M	Awaiting Material Receipt: From the time the necessary purchase order is issued until receipt of the associated material on the local receiving dock, the work order resides in this status. The responsible buyer enters this code upon the purchase order issuance. If "06" is not employed within the system, the planner enters this code upon issuing a PO request.
07/FP	Awaiting Further Planning: Once materials have been received, there may still be some planning to be completed. Depending on

the local network, either a receiving clerk or a planner should enter this code upon material receipt.

08-11/D Awaiting Required Downtime: This is a series of status codes indicating the type of downtime required to complete the work order. Downtime backlog must be separated from that which can be performed at any time (even if equipment is operating).

08/DW Awaiting Weekend Downtime: Most processes have at least some equipment down each weekend.

09/DP Awaiting Programmed Downtime: Many processes have elongated schedules (month, quarter, and year) for the operation of each line, with predetermined opportunities to relieve downtime backlog.

10/DS Awaiting "No Production Scheduled": In addition to the programmed shutdowns, equipment becomes available as a result of week-to-week production scheduling. Planners need to make themselves aware of these additional opportunities to relieve downtime backlog.

11/DO Awaiting Downtime Window: This code is for the short duration jobs requiring only an hour or two of downtime. They can often be completed during unscheduled downtime that occurs for other reasons. If planners will take a moment of additional effort to categorize downtime work orders into these four groupings, the computer can then be used to retrieve expeditiously backlog applicable to each opportunity presented.

12/R Ready to be Scheduled: Planners are responsible for getting work orders to this status as rapidly as possible. There are no encumbrances. If labor is available, the job can be scheduled.

13/FI Ready for Fill-in Assignment: These work orders are also ready to be scheduled. However, they are ideal fill-in jobs for mechanics with some form of fixed assignment. They can be readily started, dropped and resumed between more pressing demands of their fixed assignments.

14/S Scheduled: This code is used to identify jobs still open but which are on the schedule (forecast or plan depending on local terminology) for the current or following week.

15&16 Contingently Completed: Two codes are used for special control purposes. All maintenance work is complete and the jobs no longer show as crew backlog.

15/CM Completed with Materials or Rebuilding Pending: In some situations it may be desirable to charge some replacement parts not yet received to the otherwise completed work order. Possibly the removed assembly must be sent outside for rebuilding and the system calls for capturing the rebuild charges to the same work order as the "remove and replace" efforts.

16/CP   Completed, Pending Print Revisions: "As installed" prints must not be allowed to become outdated. This can easily happen, and all too often does. Even if maintenance properly marks (red lines) their working copies and forwards them to engineering for revision, an indefinite backlog frequently occurs. This backlog of print revisions can have serious adverse affect on maintenance effectiveness. It must be controlled. Status code-15 provides the necessary management awareness.

17/C   Closed: Moved or ready to be moved in to equipment history.

Accumulating Maintenance History

The last function of the WO control system is documenting the work that was actually performed during work execution. Labor and materials are charged to the work order through the time entry and stores systems. Work

Table 6-7
Trade/Skill/Unit Codes

Num. Code	Trade or Skill	Alternate Alpha Code
2	Millwright	MW
3	Carpentry	C
4	Painting	P
5	Masonry	MA
6	Pipefitter	PF
7	Electrical	E
8	Sheet Metal Worker	SM
9	Welding	W
10	Janitorial	J
11	Roads and Grounds	L
12	Air Conditioning	AC
13	Heating	HT
14	Automotive	AT
15	Instrumentation	I
16	Machine Tools	MT

Table 6-8
Work Order Classification (Type) & Purpose Codes

Code	Work Type Example
1	Breakdown/emergency maintenance
2	General repair maintenance
3	Preventive maintenance inspection
4	Corrective repair (from PM)
5	Salvage, recondition, make for stock
6	Scheduled shutdown work
7	Major outage or annual shutdown
8	Standing work orders
9	Outside charges
10	Project work
11	Alterations, modifications, rearrangements, etc.
Code	**Work Purpose**
A	Air pollution control
C	Operator convenience
E	Energy conservation
G	Economic gain
O	OSHA
S	Safety (Other than OSHA)
W	Water pollution control

performed, components affected, condition of components and cause of failure are key elements for the accumulation of equipment maintenance history from the work order. Additionally, they provide valuable information for conducting failure analysis. Reasons for Outage/Failure Code (Equipment Failure by Probable Cause) are provided in Table 6-9.

As a final note, in particularly large plants, it is advisable to develop a coding system for locations where the work is to be performed. The coding system may require as many as three or four segments to the location field. When a plant site has multiple production buildings, each building should

Table 6-9
Equipment Failure by Probable Cause Codes

Code	Probable Cause of Failure[*]
1	Improper operation or start-up
2	Improper installation or repair
3	Inadequate PM
4	Design, material, or corrosion problem
5	Lubrication or cooling problem
6	Out of balance or alignment
7	Dirty or plugged up
8	Normal wear
9	Power failure
10	Off spec. raw or packaging material
11	Other

[*]*Determined by assigned technician or other party if "root cause analysis" is performed. In some situations, agreement is reached with internal customer.*

Table 6-10
Action Taken Entries (Verb of Description)

Adjust	Escort Contractor	Remount
Align	Fabricate	Remove
Anchor	Fit	Remove and Install
Assemble	Grind	Remove and Replace
Assist	Ground	Repack
Attach	Inspect	Repair
Attend	Install	Repair and Replace
Balance	Insulate	Replace
Bend	Letter	Reroute
Bleed	Loosen	Reset
Braze/Solder	Lubricate	Retile

Table 6-10
Continued

Buff	Machine	Review
Calibrate	Modify	Roto-Root
Change	Mount	Safety Stripe
Change Over	Move	Set
Charge	No Action Needed	Setup
Check	Open/Close	Shear
Check Heat	Overhaul	Splice
Clean	Pack	Stone
Connect/Disconnect	Paint	Straighten
Cut/Sew	Patch	Strip
Deburr	Plate	Test
Delayed for Material	Predictive Routine	Tighten
Deliver	Preventive Routine	Tow
Disassemble	Rebuild	Troubleshoot
Disconnect	Recharge	Unplug
Drain and Fill	Relamp	Vibration Monitor
Drill/Ream/Tap	Relocate	Weld/Burn

Table 6-11
Condition Codes (Adjective of Description)

C01	Air locked	C12	Failed	C23	Overheated
C02	Bent	C13	Fallen	C24	Plugged
C03	Broken	C14	Faulted	C25	Scheduled PM
C04	Burnt	C15	Ground	C26	Shorted
C05	Burst	C16	Hot	C27	Tight
C06	Calibration Required	C17	Jammed	C28	Twisted

Table 6-11
Continued

C07	Cold	C18	Leaking	C29	Vibrating
C08	Contaminated	C19	Loose	C30	Wet
C09	Corroded	C20	Lube Required	C31	Worn
C10	Dirty	C21	Misaligned		
C11	Eroded	C22	Noisy		

Table 6-12
Component (Noun of Description)

AC Motor	Carrier	Guide	Roller
Agitator/Mixer	Camera	Handrail	Roof
Air Compressor	Clamp/Pad/ Pin/Fork	Hasp	Rotameter
Air Conditioner	Clutch	Head	Rotating Assembly / Impeller
Air System	Coating Pans	Heat Exchanger	Rotary Joint
Alarm	Coil	Heater/Radiator	Rotor
Amplifier	Column	Hoist/Boom	Safety
Analyzer	Cabinet	Hood	Saveall
Antenna	Cable-Chain	Hose/Line	Scales
Armature	Commode	Housing/Frame	SCR Drive
Auger	Commutator	Hydraulic System / Equipment	Seal
AC Motor	Compressor	Impeller Wear Rings	Seat
Agitator/Mixer	Computer	Impeller	Sewer
Air Compressor	Condenser	Impulse Line	Shaft/ Journal/Axle

Table 6-12
Continued

Air Conditioner	Conduit	Indicator	Shaker -Coal Car
Air System	Connector	Fuse	Shear
Base	Contacts	Gasket	Sheave
Base	Container	Gauge	Sight Glass
Base	Controller	Gear	Rectifier
Base	Controls/Elect. or Hydro	Gear Box	Reducer / Gear Box / Transmission
Base	Conveyor	Generator/ Alternator	Reel
Base	Cord	Instrument Misc.	Refractory - Boiler
Base	Count Rate Meter	Insulation	Refrigerator/ Refrigeration
Base	Coupling	Integrator	Regulator
Base	Cover	Jack	Relay
Base	Crane	Joint	Reorder
Base	Cup	Jordan/Refiner	Resistor
Base	Cylinder	Jumper	Rheostat
Base	DC Motor	Key	Ring
Base	Decade	Ladder	Road
Base	Detector	Lamp	Sign
Base	Diaphragm	Lawn	Silica Gel
Base	Dies	Lead	Siphon
Base	Differentiate	Lead Cell	Sleeve
Base	Disconnect	Level indicator	Slide Wire
Base	Door	Lighting	Slitter
Base	Door Closer	Linkage	Solenoid

Table 6-12
Continued

Base	Door Operator	Lock	Source
Base	Drain	Lower Block / Spreader Bar	Spring
Base	Drinking Fountain	LP System	Sprocket
Base	Drive	Magnet	Starter/ Disconnect
Base	Drum	Materials/Parts	Stator
Base	Electric Brake	Mechanical Misc.	Steam
Base	Electric Hoist	Meter/ Instrument	Steel
Base	Electrode	Misc. Equipment	Steering
Battery	Element	Motor	Strainer
Bearing	Elevator/ Lowerator	Mounting	Strips
Bellows	Equipment	Nozzle	Surface
Belt	Exchanger	Nut	Switch
Bench/Shelf/ Locker	Fan-Blower	O-Ring	Switch Gear/ Circuit Breaker
Blade	Fan-Housing	Oil	Tach Generator
Blow-out Plug	Faucet	Overload/Fuse	Tank/Chest
Boiler	Feeder	Packing	Terminal
Bolts/Pins/Screws	Fence	Panel	Thermal Couple
Bonnet	Filter/Screen	Parking Lot	Thermal System
Box	Fire Protection System	Pen-Pointer	Thermostat
Bracket/Hanger	Fitting	Picture-Poster	Timer
Brake	Fixture	Piping/Tubing	Tire

Table 6-12
Continued

Breaker	Flange	Piston/Plunger	Tool
Brushes/Holder	Fleet Vehicles	Plug	Tractor
Bucket	Flight	Plumbing	Transformer
Building	Float	Positioner	Transistor
Bushing/Shim	Floor	Power Supply	Transmitter
Camera	Fluid	Precipitator	Transporter
Cable-Computer	Fly Wheel	Press	Trap
Cable-Electrical	Forklift	Pressure Switch	Trash
Capacitor	Form/Staging	Probe	Trim
Capillary	Fountain	Production	Tube Sheet
Cart	Frame	Programmable Controller	Tube
Casing	Freon	Pulley/Sheave	Tubing-Fittings
Casing Wear Rings	Furniture/Office Equipment	Pulper/Beater	Turbine
CCTV	Gasket	Pump	Union
Chain	Gauge	Punch	Unitrol Disconnect
Chamber	Gland	Rack	Unwind
Chart	Glass-Gauge	Reboiler	Vacuum Tube
Chemical	Governor	Receiver	Valve Fittings
Chip	Grate	Receptacle	Wall
Cabinet	Grate Drive	Recorder	Water Cooler
Cable-Chain	Grinder	Rod	Watt Hour Meter
Cable-Computer	Grounds	Roll	Wear Rings
Cable-Electrical	Guard	Roll Wagon	Web Aligner

be numbered or abbreviated. General location within structures can be coded by geographic reference or production zones or both. For example, the location for work to be performed in the Final Assembly Building, on the rolling stock (large) assembly line # 2 in the northwest corner of the building might be coded as FAB.NW.RSL2.

6.1.4 Sequence of Planning

The planning and scheduling function is the center from which all maintenance activity is coordinated. While planning and scheduling are closely related, they are distinct functions:

- Planning (how to do the job): Planning is advanced preparation of selected jobs so they can be executed in an efficient and effective manner during job execution that takes place at a future date. It is a process of detailed analysis to determine and describe the work to be performed, task sequence and methodology, plus identification of required resources, including skills, crew size, labor-hours, spare parts and materials, special tools and equipment. It also includes an estimate of total cost and encompasses essential preparatory and restart efforts of operations as well as maintenance.
- Scheduling (when to do the job): Scheduling is the process by which required resources are allocated to specific jobs at a time the internal customer can make the associated equipment or job site accessible. Accordingly, the preferred reference is "Scheduling and Coordination." It is the *marketing arm* of a successful maintenance management installation.

Considered together, these two elements represent "Work Preparation." Their importance and absolute essentialness are predicated on the principle "Maintenance management will achieve best results when each worker is given a definite task to be performed in a definite time and in a definite manner." The supervisor, the worker and their colleagues should each know what is expected—including the goal and target time for completion of each job. The planning process begins with the following considerations:

- Control minor maintenance work activities within the plant work control system.
- Determine the level of detail necessary to accomplish maintenance tasks and troubleshooting.
- Use maintenance history in planning corrective maintenance and repetitive job tasks.

- Identify needed support to perform maintenance.
- Prepare and assemble a maintenance work package. Related follow-on to this includes:
  - procedures and work package approval
  - work package closeout and maintenance history update.

A system of planning, scheduling, coordinating, executing and closing out maintenance work activities should be clearly defined based upon a plant Standard Operating Procedure (SOP), which should consist of five interrelated processes applicable to each maintenance job. These processes are:

1. Plan Maintenance Job. Identify the scope of a needed maintenance job. Produce a maintenance job plan. Determine maintenance job planning category, priority, and safety concerns. Identify and procure materials, and identify other maintenance task resources. Prepare the maintenance job package.
2. Schedule Maintenance Job. Calculate estimated start date and project resources for the maintenance job. Schedule and commit required resources and special tools/equipment items to allow performance of all maintenance tasks within the maintenance job.
3. Execute Maintenance Job. Initiate and perform a maintenance job and collect job information as defined in the maintenance job package.
4. Perform Post-maintenance Test. Verify facilities and equipment items fulfill their design functions when returned to service after execution of a maintenance job.
5. Complete Maintenance Job. Perform maintenance job closeout to include completion of all documentation contained in the maintenance job package to ensure historical information is captured.

6.1.4.1 Job Plan Level of Detail

Actually, before even considering the level of detail, the planner must determine or decide if this Work Order should be planned. Planning thoroughness increases as the installation matures. Any factors that might delay or hinder effective job completion should be anticipated and provided for in the "planning package." Planning completeness does vary with job repetitiveness. If the job is repetitive in nature, the planner can afford to invest more time since his or her efforts will yield repetitive returns in months and years to come. "Planning efforts will yield residuals."

Conversely, if the job is clearly a one-time effort, planning should be taken only to the extent that the effort will result in a completed, reliable and timely executed maintenance action. Longer duration jobs will also benefit proportionally from more detailed planning. Additionally, it is often the case that components of even nonrepetitive planning packages can be reused if the job is effectively planned in "building blocks."

These are prime reasons for providing sufficient planning capacity in early phases of program installation. When there is enough planning capacity to avoid the "hurry up" syndrome, the preferred approach is to prepare a very thorough planning package for any repetitive job encountered. In this manner, planner reference libraries are established more rapidly and planner workload diminishes as repetitive effort diminishes through the expansion and use of reference libraries.

The appropriate detail or degree of planning is often open to question. Unnecessarily detailed planning on simple jobs can be more work and more expensive than can be justified. It is normally concluded that the smaller the job, the less planning required. However, the actual tendency is to under-plan normal-size jobs rather than to over-plan the simple ones. Even a small one-hour job that is mistakenly made available with some essential information or materials missing can cause considerable loss of time due to unnecessary travel and/or reassignment. Jobs of medium to small size are frequently neglected relative to planning detail.

The eventual answer to the original question of whether the work order should be planned is, "Plan all work orders whose performance can benefit from planning." At the same time, planning should not become an insult to a tradesperson's ability. It pays to have a good familiarity with the skill levels of the tradesperson whose work you are planning.

The question of planning detail applies primarily in early implementation phases of the planning and scheduling functions, when there simply is insufficient time to plan all jobs effectively. This situation diminishes as planner references are developed and reference files of preplanned work packages are built for repetitive jobs. Planners must avoid getting bogged down in unscheduled and emergency work, if any other maintenance planning is going to be completed. Following are derived definitions applicable to level of detail progression in job planning:

Simple
Every Job Plan/Work Order will have reasonable detail in its description and scope of work
Safety Issues will be identified
Labor estimates and downtime will be established
Skills will be determined and assigned

Priorities have been established with the requester/customer

Job Plan follow-up comments are solicited

Communication with requester/customer prior to starting work and after work completed.

Average

Everything in a simple job plan will be expected as well as the following:

Detailed JSA's, increased detail in the disassembly, assembly, installation, corrective repair steps

Specifications are included. Ex. Torque values, clearances, wear limits, etc.

Materials are allocated and received prior to work starting

Special tools and equipment identified

Work area housekeeping and tool returns

More details in follow-up comments expected for improvement of process.

Advanced

Everything in a simple and average job plan will be expected as well as the following:

Engineering drawings included as the complexity of the job plan warrants

Schematics supplied as part of the job plan as necessary

Testing requirements and Quality controls detailed

Special needs and hazards are identified within the job plan.

Not all job plans will require the complete details of the advanced. However, all work orders should have the fundamental elements of the Job Plan Template.

In early phases of Planning and Scheduling implementation, a two-, four-, or even eight-hour cutoff point is often established for the magnitude of jobs that will be planned. As indicated, the selected cutoff point should be progressively reduced as the planning and scheduling function matures. Generally, larger jobs are planned first on the theory that there is a greater likelihood of delay and conflicts and therefore a greater opportunity for positive impact by planning. Conversely, delays encountered in smaller jobs have a more dramatic percentage impact.

The Pareto Principle

The Pareto principle, often referred to as the 80/20 rule, states that the most significant items in a given group normally represent a relatively small portion of the total items in the group. The wisdom of focused and well thought out effort is apparent and indicates that effort to control the "vital few" can bring results out of proportion to the efforts expended. Figure 6-4 graphically illustrates the Pareto Principle as it applies to the work planning function.

6.1.4.2 The Work/Job Package

Table 6-13 depicts the contents of a full or complex Planned Work Package. Not all of the nine elements are required for every work package. The remainder of this section outlines the processes and considerations for development of the full work package.

There are conditions or practices that can enhance the work planning process. Among them are:

1. Complete equipment repair history
2. Thorough PM inspections

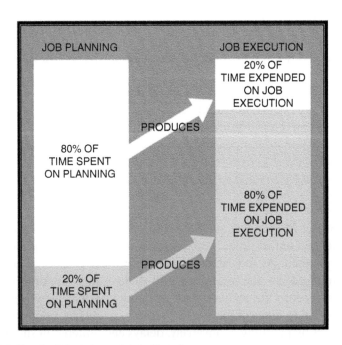

Figure 6-4 Pareto Principle and Job Planning

Table 6-13
Contents of the Work Package

Work/Job Package	
1	Work Order/Job Plan (includes resource requirements)
2	Technical Manual (if applicable—procedures, drawings, etc.)
3	Pre-Test/Pre-Maintenance Checks (if applicable)
4	Bill of Materials
5	Drawings/Sketches/Photographs
6	Step-by-Step Procedures
7	Time—By Step, By Allowance
8	Re-Test Requirements/Procedures
9	Post-Maintenance Notification and Reporting Requirements

3. Timely reporting of potential problems by production
4. Thorough failure analysis by Reliability Engineering
5. Open dialogue with operators on troublesome equipment
6. Supervisor awareness of impending problems
7. A planned component replacement program
8. Existence of in-house overhaul and rebuild capabilities
9. High quality of work by trade personnel
10. Good use of repair technology

Work/Job Package Development

Long- and short-range job planning falls into six phases:

1. Initial Job Screening
   a. Assisted by the Maintenance Administrator, the Planner receives *all* requests with the exception of emergencies for maintenance work (those emergencies that must be performed on the same day as requested will be recorded and passed immediately to the Maintenance Supervisor).
   b. Requests are reviewed and screened for completeness, accuracy and necessity:
      i. Clear description of request
      ii. All requestor fields filled in with valid codes

    iii. Priority and requested completions are realistic and provide practical lead time

    iv. Authorization is proper
- Develops preliminary estimated if required to obtain approval
- Obtains Engineering approval for all alteration and modification requests

    v. The requested work is needed:
- Has it already been requested?
- Does it need to be accomplished? If so, does it need to be accomplished at this time?

If questioned, the issue is resolved with the requesting Department or referred to the Maintenance Supervisor.

2. Analysis of Job Requirements. The Planner examines the job to be performed and determines the best way to accomplish the work—consulting with the requestor and/or the Maintenance Supervisor as appropriate. In determining job requirements, the Planner:
   a. Determines the required level of planning
      i. Does this job warrant detailed planning or should it bypass the planning process completely?
      ii. Is the effort and cost worth the value to be gained?
   b. Visits the job site and analyzes the job in the field. One-third of the Planner's day should be spent in the field:
      i. Conferring with the requestor
      ii. Clarifying the request and refining the description:
         - Where the job is located (machine or location)
         - What needs to be done (job content)
         - Start and finish points (job scope)
         - Finalize priority
   c. Visualizing job execution and outlining the requirements
      i. Mentally go through and record the steps necessary to execute the job
      ii. Prepare sketches or take pictures to clarify intent of the Work Order for assigned mechanics or simply as reference for self during detailed planning
      iii. Take necessary measurements (exactly)
      iv. Determine required conditions. Must this job be coordinated with Production?
         - Must equipment be down (major or minor?)
         - Define involved control loops

- Will other equipment or adjacent areas be impacted by performance of this job?
- Check for safety hazards.

3. Job Research. The remainder of the planning cycle is normally completed at the planner workstation (two-thirds of the Planner day). During job research, the Planner:

   a. Uses Equipment History to determine if the job has been previously performed:

      i. When was the last time?

      ii. Is it excessively repetitive? If so, consider if anything can be done to avoid recurrence.
         - Is this the best solution to the problem?
         - Consider the alternative approaches
            ○ Should additional work be performed in the interest of a more permanent solution?
            ○ Repair/Replace
            ○ Make/Buy
            ○ Consider engineering assistance
         - Be conscious of alternate plans for the involved equipment.

   b. Refers to planner libraries and to the file of planned jobs to determine if the job or portions thereof have been previously planned

      i. Use what you can, avoid redundant effort

      ii. Reference the procedures file
         - Determine safety requirements
         - Identify necessary tag outs
         - Identify necessary safety inspections, fire watches, and standby positions associated with ladder safety, vessel entry, etc.

      iii. Safety must always be a top priority of job planning

   c. Talks to other functions with involvement or potential input in the job

      i. Is engineering assistance required?

4. Detailed Job Planning. During detailed job planning, the Planner details and sequences job requirements:

   a. Selects and describes the best way to perform the job

   b. Determines and sequences the job by specific and logical tasks or steps

      i. Identifies task dependencies and considers application of PERT or CPM network analysis to facilitate the planning of complex jobs (refer to Appendix D)

      ii. Determines required skill sets for each task (trade and skill level)

      iii. Prepares cross work orders to other groups as required and necessitated by the CMMS in place

      iv. Determines the best method for job performance (the best way to do it)

    c. Determines resource requirements

        i. Establish the required crew size and labor-hours for each task of the job sequence
- Estimate or apply available benchmarks
- Apply job preparation, travel and PF&D allowances
- Determine if extra travel or job prep is needed

        ii. List determinable materials, parts, and special tools required
- Prepare the Bill of Materials
- Establish the acquisition plan
    - Determine what items are in stock and reserve them
    - Source those items which must be direct ordered (Purchasing responsibility)
    - Prepare acquisition documents
        - Stores Requisition for items in authorized inventory
        - Purchase Requisition/Order for direct purchases
        - Work Order for in-house fabrication
        - Purchase Order with MWO reference for contractors and outside equipment rental
    - Don't forget special tools and equipment
        - Ladders and Scaffolding
        - Rigging

       iii. Determine equipment and external resource needs

       iv. Consider disposal issues (expense, time, special handling)

       v. Estimate total cost in terms of labor, material and external charges

      vi. Coordinate and expedite necessary authorizations based on final cost estimate
- Operational
- Financial
- Engineering

5. Job Preparation. During job preparation, the Planner assembles the planned job package. This package for any given job contains documentation of all planning effort. Given the data contained within the package, coupled with a thorough verbal exchange between Planner and Maintenance Group Supervisor, followed by similar exchange between Supervisor and assigned tradesperons, nothing should be lost between strategic planning and tactical execution of the plan.

    a. Complete and detailed Work Order

    b. Job plan detail by task

        i. Step by step procedures

       ii. Site Set-Down Plan (if a significant tear down)

   c. Labor deployment plan by trade and skill
      i. Labor-hour estimates
      ii. Consider contract as well as in-house resources
      iii. Consider the use of the GANTT bar chart to help convey task sequencing to assigned crews
      iv. Maximize pre-shutdown fabrication and other preparation
   d. Bill of Material
   e. Acquisition Plan
      i. Authorized Inventory vs. Direct Purchase
         • Including availability, commitment and staging location
         • Make or Buy decision (cost and cost of time considerations)
      ii. Required Permits, clearances and Tag Outs to the point feasible and safe (final steps must be taken and verified by the responsible mechanic and equipment operator)
      iii. Prints, sketches, digital photographs, special procedures, specifications, sizes, tolerances and other references which the assigned crew is likely to have need of

As appropriate, the assembled package is reviewed with the Maintenance Supervisor and the Requestor. The Planner then holds the Planned Job Package for necessary procurements.

  6. Procurement. Within the Procurement Process:
   a. Purchasing sources all materials requiring direct purchase, obtains necessary competitive bids, obtains delivery dates, and cuts associated purchase orders
      i. Monitors and expedites materials/parts delivery
   b. Receiving all maintenance materials whether for direct purchases or stock replenishments
   c. On basis of the Weekly Master Schedule, the storeroom picks, kits, stages and secures those scheduled job items available from Stores
      i. Documents all requisitions via the CMMS
      ii. Stocks and maintains the Maintenance Storeroom including regular cycle counting to assure continuing accuracy of inventory

The job is released for scheduling only when all required resources (other than labor resources) are on-hand.

Job Plan Requirements

Following is an outline of items that should be included within the job plan text or as part of the job package when the work is scheduled. *Not* all job plans require all of the listed steps.

1. Purpose of the work order
   a. This could include what has taken place already, such as a follow-up to a repair of breakdown.
   b. This could include the initial write-up and who made the observation.
2. Individuals/Departments to be notified
3. Safety-Related Issues.
   a. Lockout/Zero Energy State procedures. (These can be included in the text or as an attached procedure.)
   b. Special personal protective equipment (PPE) requirements
   c. Required Permits
      i. Confined Space
      ii. Hot Work
   d. Special Hazards (chemical or other)
      i. Any solvents, chemicals or coatings that may be required
   e. Safety Equipment
      i. Barricades, fixed or portable (tape or cones)
      ii. Signs, Men at Work, Overhead Work
      iii. Safety harnesses, electrically insulated gloves, etc.
      iv. Crane rail stops
4. Special Tools and Equipment
   a. Cranes
   b. Man-lifts
   c. Special Hand Tools
5. Equipment Drawings
   a. Engineering
   b. Sketches
6. Operation and Maintenance Manual Procedures
   a. These can be included in the text or as an attachment, both electronically or hard copy.
7. Job step sequenced procedures.
   a. These can be included in the text or as an attached procedure.
   b. May list in the text of the step the storeroom part numbers that are not anticipated or ordered but may be required.
8. Specifications and tolerances.
   a. Torque requirements
   b. Wear Limits
9. Special Instructions
   a. Special access sequence, etc.
10. Reference to trade interaction
    a. Electricians need to lockout at a special time
    b. Riggers will be arriving at a certain time.

11. Testing Requirements
    a. Quality checks
12. Clear Lockouts and/or Permits
13. Notify Individuals/Departments work completed and equipment has been returned to service
14. Cleanup and Housekeeping of work area
    a. Return special tools to crib, shop area or where necessary
    b. Disposition of replaced parts (rebuild—in-house or vendor, trash, hazardous waste, reliability engineering, etc.)
    c. Dispose of waste oil
15. Complete paperwork
    a. Input time against the work order
    b. Make comments on work order
        i. Follow up work required
        ii. Improvements to the job plan steps
        iii. Addition parts that may have been used that need to be added to the equipment bill of materials
    c. Turn the completed work order into supervisor/foreman

6.1.4.3 Estimating and Work Measurement

A necessary ingredient of any operating control system is the generation of quantitative standard information against which to measure and assess performance in the execution of work; to identify areas for corrective action; to measure progress towards goals; and to aid in the decision-making process. The objective of work measurement is to generate this standard information.

Realistic labor estimates are an essential part of a planned maintenance program. There is no effective way of matching workloads against available labor resources without measurement, and it is hard to make realistic promises when taking equipment out of production. Estimates are also essential when determining what the correct manning should be for each grade of labor and what level of crew performance is being attained.

Familiarity with Maintenance Jobs and with Plant Equipment. Some of this familiarity the planner brings to the job, some is acquired on the job. There is no substitute for task knowledge when it comes to estimating. A trade background is an ideal starting point; it enables Planners to "visualize" actually doing the job themselves. The next best thing is actually observing what is involved as jobs are worked and, when this is not possible, "talking through" a job with a maintenance tradesperson or supervisor who is familiar with it.

Becoming involved in the job, and being seen as involved, has tremendous advantages. Involvement expands the planner's knowledge of the plant and its equipment. Possibly more important, it builds credibility between the maintenance tradesperson and supervisor for the planner as well as what the planner is doing.

Levels of Maintenance Work Measurement Methodology. Several forms of work measurement, all with varying levels of precision, can be used in the development of job estimates:

1. Supervisor/planner estimates
2. Historical averages
3. Published job estimating tables (construction trades)
4. Adjusted estimates or averages (based on work sampling during a base period)
5. Analytical estimating
6. Time study
7. Predetermined times
8. Predetermined time formulas

The method(s) used depend significantly on the focus of the installation, i.e., Performance Measurement or Schedule Compliance. A Performance Measurement focus (Standard Time ÷ Actual Time = Performance %) requires a more precise form of work measurement.

A Schedule Compliance focus emphasizes early planning and scheduling using less precise forms of work measurement. Analytical Estimating is the preferred form.

Analytical Estimating. Analyzing and estimating maintenance work seems difficult at first because there are nearly always elements that are unpredictable. Normally however, the unpredictable elements do not constitute the whole job and are quite often only a minor part of all that has to be done.

The purpose of analytical estimating is to develop reasonably accurate and consistent time estimates. The technique is relatively straightforward and is quickly mastered. It is based on the following principles:

1. For persons who have had practical experience doing the work, it is easy to visualize and set a time, for simple, short duration jobs.
2. Long, complex jobs cannot be estimated as a whole.
3. Pinpoint accuracy in estimating is not justified or achievable since all the variables in maintenance work cannot be known until *after* the job is completed.
4. Estimating is easier and more accurate when the job is broken down into separate elements, or steps, and each element is estimated separately.

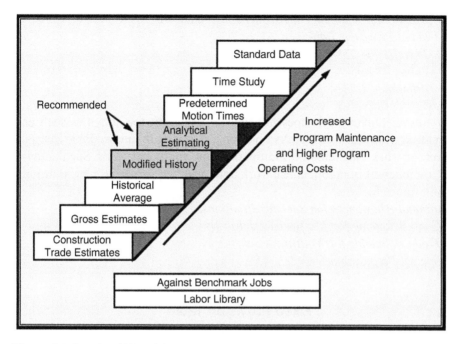

Figure 6-5 Levels of Work Measurement

5. All jobs can be broken down into the following sequences:
   λ Getting ready and receiving instructions for doing the job. This includes receiving job instructions from the supervisor; collecting personal tools together; getting parts from storeroom; collecting special tools and equipment.
   Travel to job site.
   λ Diagnose/assess problem.
   λ Shut down machine before starting work. This includes finding the line supervisor; stopping the machine using correct sequence; locking out procedure; receiving production input on problem.
   v Partial or total disassembly to get to the problem area.
   λ Determine full extent of problem.
   λ Identify replacement parts needed. Look up in parts list and catalogs. Collect from storeroom or special order.
   v Reassemble machine using replacement parts as necessary.
   λ Check, test and clean up job site and put away tools. Travel back to shop.
   λ Report on job and return special tools and equipment.
   δ Allowances.

Key to Estimate Source for the sequences on page 163:
λ   *-Fixed Provision Table*
     -Travel Time Table
ν   *-Labor Library*
δ   *-Allowance Table*

Allowances. Direct work does not include provision for activities such as preparation, authorized breaks and wash up, fatigue, unavoidable delays, travel, or work balancing on multiperson and multitrade jobs. Such activities are inherent in maintenance work and must be provided for by addition of allowance factors.

Additional Allowances for Specific Situations:
   Crew Balancing for Multiperson Jobs=3%
   Trade Balancing for Multitrade Jobs=2%
   Special Preparation for ?????? =5%

Table 6-14
Fixed Provision Table

Element	Simple	Average	Complex
Job paper work	Normal—work order, TDC, etc.	Preventive Maintenance, checklists, etc.	Feedback Failure Codes
Instructions	Up to 3 minutes	Approximately 5 minutes	Approximately 10 minutes
Tools	Normal— carried on belt	Obtain 3 or 4 from personal toolbox	Special—ladder, drill motor, torque wrench, etc.
Safety	Normal	Permits	Special precautions—rope off area, protective breathing, etc.
Parts	None	1 to 5 different items	6 or more different items
Sketches/ Drawings	None	Simple	Complex— Blueprint, electrical schematics, etc.
Allowed hours per man	0.08	0.17	0.33

Table 6-15
Travel Time Table

From Maintenance Shop To	Round Trip Hours	Allowed Hours		
		Simple	Average	Complex
Area A	0.5	0.5	1	1.5
Area B	0.4	0.4	0.8	1.0
Area C	0.3	0.3	0.6	0.6
Area D	0.2	0.2	0.4	0.8
All Other Facility Areas	0.1	0.1	0.2	0.3

Comparative Time Estimating (by making sure everybody compares to the same known): Absolute accuracy is neither possible nor necessary for maintaining an acceptable level of efficiency and control in maintenance. What is essential is *consistency*.

It is in the nature of human mental processes to estimate by comparison. We compare the unknown with the known and then estimate the degree of similarity (identical, larger, smaller . . . how much?). The tendency is to do

Table 6-16
Allowance for Personal, Fatigue and Delay (P, F & D)

Category	Percent	Minutes	Light	Average	Heavy
Personal Time (Breaks & Clean-Up)	5	30	30	30	30
Unavoidable Delay	5	20	20	20	20
Fatigue	5 to 10		10	20	30
Total:	15 to 20		60	70	80
Percent (Total/ 480-Total)			15%	17%	20%

Add the light work total (15 %) to preparation and travel time. Add the appropriate total to direct work (15, 17 or 20 %).

this automatically and very subjectively. Some people have an almost infallible sense of comparative size; others find it hard to come within a mile. The basis for consistent estimating is for all planners to have access to and use the same library or catalog of job estimates. There are four basic methods for structuring comparative job estimate files for maintenance work:

1. Systematic files
   a. By skill
   b. By nature of work
   c. By crew size
   d. By labor-hours
2. Catalog of standard data for building job estimates
3. Catalog of Benchmark Jobs
4. Labor Library

Current state-of-the-art rests between the last two methods.

Regardless of the comparative approach used, the basic form of work measurement must still be decided. The standards (or estimates) to be loaded into the comparative catalog must first be developed.

The technique of comparative estimating involves the *comparison* of jobs with those in the library, not the *matching* of jobs. This distinction is important because in the former case, a few hundred carefully selected jobs will enable an experienced Planner to produce consistent estimates for most maintenance work. In the latter case, many thousands of jobs would be needed to get the same result.

To make a comparative estimate, the planner must (1) define the scope of the job that is to be done and (2) prescribe the method to be used. The planner will need a good knowledge of the process and equipment, and will often have to visit the job site and talk with the operations and maintenance supervisors as well.

The planner's next step is to turn to the appropriate section in the benchmark library to try to find a benchmark with a description similar to the job for which an estimate is required. The planner must then make a judgment based on his or her own mental comparison of what is involved in doing the benchmark job (for which a time is available) and what will probably be involved in doing the job for which an estimate is needed. Final judgment is based on four basic decisions:

1. Is the new job bigger or smaller than the benchmark job it is compared with?
2. Is the difference so small that it will remain in the same time interval?
3. Will it fall into one of the other intervals, either above or below it?

4. Does the new job fall into the next time interval above or below, or two intervals above or below?

Comparative estimating still involves subjective judgment on the part of planners, but the only choice that they need to make is between one time interval and the next. Comparison with library jobs of known duration greatly reduces guesswork on the part of the planner. Maintaining the reference library is a central function requiring the assembly of contributions from all planners to cover the various classes of equipment. In this way, a uniform structure for job estimates can be established and controlled.

6.1.4.4 Planning Aids

The informational sources required for efficient maintenance planning fall into several categories:

Material Libraries filed by each unit of equipment on which maintenance work is performed. The material library supports development of the bill of materials and the material cost estimate.

Labor Libraries filed in some conveniently retrievable manner—normally by unit of equipment, specific type of skill required or by job code. The labor library supports development of job step sequence and the labor cost estimate.

Equipment Records recording all pertinent data for equipment such as installation data, make and model, serial number, vendor capacity, etc. Equipment Records should not be confused with Equipment History of actual repairs made to the equipment.

Prints, Drawings and Sketches as installed/as modified.

Purchasing/Stores Catalogs providing pertinent information not captured in the material libraries. Even where a materials library is in place, some form of stores catalog or vendor catalogs were used to develop it.

Standard Operating Procedures (SOPs) that can be included in planning packages without repetitive documentation effort. Such procedures include safety, lockout, troubleshooting sequences, etc.

Labor Estimating System providing basic data for building job estimates. Even where a labor library is in place, some form of estimating system was used to develop it.

Planning Package File of previously developed packages for repetitive jobs, enabling repeat usage with minimal repeat planner effort.

Additional reference sources:

1. Catalogs
2. Service Manuals
3. Parts List
4. Storeroom Catalogs

5. Estimating Manual
6. Engineering Files
7. Equipment History
8. Procedural Files
9. Experience
10. Supervisors
11. Reliability Engineers
12. Mechanics
13. Operators
14. Feedback

Labor, Materials and Tool Libraries (Planning Aids). Some planning functions vary each time they are carried out. Others follow the same pattern for each group of identical machines. It is therefore possible to simplify some of the planning processes for machine repair and overhaul by *classifying* identical groups of machinery and then *building* libraries of preplanned work element sequences and bills of materials for each class.

The basic concept of these libraries is to take each type of equipment, class by class, and to establish the job sequences necessary to take it completely apart and then put it back together again. Appendix C, Exhibit C10, provides a filled-in example of a Labor/Materials/Tools Library sheet that shows at a glance for each element:

1. The trades and estimated hours to do the job.
2. The parts involved.
3. The stockroom number of each part.
4. The manufacturer's ID for each part.
5. All special tool and equipment requirements.

Some or all of these sequences are involved in every job done on given equipment. Some additional sequences may be required when parts are repaired rather than replaced, but once these libraries are established:

1. The planner's job is simplified.
2. Every job plan is consistent.
3. There is a good foundation for computer assistance.

A number of sources should be utilized as the situation dictates:

1. equipment history;
2. procedure files;
3. experience;
4. supervisors;
5. reliability engineers;
6. mechanics;
7. operators;
8. feedback.

6.1.5 The Role of CMMS in Maintenance Planning

Discussions of planning, coordination and scheduling of the maintenance function to this point have been addressed from the "done by hand," or without computer support, perspective. Until recently, maintenance planning was indeed accomplished without computer support. Today, however, with the proliferation of new and improved Computer Managed Maintenance Systems or Computerized Maintenance Management Systems (CMMS) and Enterprise Asset Management (EAM) software, job preparation can be accomplished far more efficiently with computer support.

With the efficacy and tailorability of today's array of information management software, it is no longer cost-effective to manage the maintenance function without computer-managed information access. Computer support is vital to the maintenance control system if cost-competitive posture is to be maintained. Only computer information processing systems that are outputting (reporting) and inputting (updating) data nearly continuously can meet the accuracy and speed required in today's maintenance environment. Based on the essential and mandatory nature of computer-managed maintenance, information availability to the Maintenance Planner, perhaps the most immediate concern is the establishment of a dialog with the IT Department.

The first step, if it has not previously been integrated, is the setup of the work order system in the (CMMS). All of the WO fields, coding tables and related standardized WO data, to include computer-generated WO number assignment and tracking must be provided to/by the software. It is important in this early stage of Maintenance Planning and Scheduling integration into CMMS that the planners and schedulers not get bogged down in the intricacies of data field design, formatting and related software setup and tailoring. Provide the baseline information and expect IT to take it from there. Periodic sit-downs with IT to evaluate progress and provision will keep the planner up-to-date as well as identify any misdirection in the CMMS work order control system.

If a CMMS is already in place, an equipment database (equipment records) should exist. The accuracy and completeness of these equipment listings (including drill-down access to a breakdown of major components) is often dubious. Elicit the support of the maintenance manager to have the equipment database validated for completeness as well as accuracy. A widely and randomly performed spot check should quickly identify whether problems exist. If the equipment listing does not exist however, the task of setting it up will be much too labor-intensive to be accomplished in-house. Look to outsourcing to a firm specializing in "CMMS Implementation" to accomplish this and similar *data-intense* efforts.

Following the validation of the equipment records database and setup of the work order system, attention should be directed toward integrating the backlog management function within the CMMS. WO System setup in CMMS will provide at least 50% of the data needed for backlog management. Providing resource information (updated weekly via the maintenance supervisor's resource availability report) as well as available work status (updated through the work order completion process) gives CMMS nearly all that it needs.

The remaining tools for facilitating the planning and scheduling function involve basic CMMS capabilities. Many CMMS and EAM vendors provide their software in modules that are capable of completely merging with all other modules or elements of their software. If a specific Planning and Scheduling module is not available, they should be able to integrate with a project management program as well as a report generator. The functionalities required include:

1. Storage and Retrieval of:
   a. Equipment History
   b. Work/Job Plans and Estimates
      i. Job Steps
      ii. Permit Requirements
      iii. Safety Precautions/Steps
   c. Current Inventory (and ability to directly requisition or allocate from inventory)
   d. Indexed Technical Manual and Drawing Files including locations
2. Computation, Execution, Reporting and Updating For:
   a. Life Cycle Cost Analysis Using Integrated Inputs From
      i. Maintenance Cost Data (labor, parts, etc.)
      ii. Purchasing Data (equipment procurement costs, trade-in value, etc.)
      iii. Accounting Data (equipment depreciation rates, etc.)
      iv. Production Data (downtime, output quality factor, output rate, etc.)
   b. Backlog Size and Composition (available and unavailable)
      i. Estimating Capability for Unavailable Backlog size
      ii. Alarm (Notification) Point Set (i.e., when available becomes less than 80% of total backlog)
   c. Backlog/Resource Balancing calculations
   d. Shutdown/Outage Correlation to Available Work (Orders)
   e. Work Order Aging including recommended actions (i.e., re-prioritizing, canceling or scheduling)

3. Automation/integration of information flow from control systems and condition-monitoring software[*]
   a. Equipment run-time and startup/shutdown cycles
   b. Real time operating conditions (temperature, pressure, flow, etc.)

There are additional benefits to be gained by utilizing the CMMS or EAM to facilitate such things as life cycle cost analysis, work measurement, labor efficiency calculations, failure analysis, maintenance optimization, PdM and CM analysis, PM scheduling and other functionalities that either the Planner or Scheduler, or both, will interface with or otherwise utilize.

6.1.6 Feedback

It is important for the Maintenance Planner to know if his planning and work packages are providing the trades with everything they need to perform the work requested on the work order. Even though human nature tends to produce or elicit criticism when things are not right, it is doubtful that criticism of the planner's efforts can produce much improvement since it usually lacks focus. As a result, it is up to the planner not only to elicit feedback regarding his or her performance, but also to give it the focus necessary to enact improvements.

A proven method to measure planning quality is through the use of post-completion feedback and critique. Measuring planning quality must be an ongoing effort. The maintenance manager or planner should hold regularly

---

[*]*Note: Integration of control system and condition monitoring data into CMMS or EAM systems is not a widely offered capability provided by CMMS or EAM software vendors at the time of this writing. However, in plants practicing Reliability-Centered Maintenance (RCM) having this feature can be considered essential. The primary purpose of CMMS is to support work management and execution, however, most preventive maintenance (PM) tasks are performed based on calendar- or meter-based schedules and condition monitoring and predictive maintenance technologies are used to analyze equipment condition data to identify potential problems before they occur. Once the need for a PM is identified and either CM or PdM identifies a problem, work orders can then be created in CMMS by keying the information in. Does not this seem to be a logical application for CMMS automation? Meter readings and inspection-point data can be collected on hand-held computers or reported by equipment control systems. Predictive maintenance analysis software can identify the corrective work required to address a potentially downward trend in an asset's performance. Alarms generated by monitoring and control system software, through integration with CMMS, can be used to generate emergency work orders.*

scheduled meetings, preferably weekly. For efficiency, the feedback meeting could be an agenda line item in the weekly finalization meeting for the maintenance schedule. Periodically the meeting might be chaired by the plant manager to enhance the importance of the meetings. The topics for the feedback portion of these meetings include critique of the recently completed schedule as well as finalization of the upcoming schedule. The critique session is the opportunity to assess planning quality by specifically addressing the questions:

1. Was the schedule successfully completed? What was schedule compliance?
2. Were any of the schedule shortfalls due to incomplete or poor planning?
   a. What was the problem?
   b. Could it have been avoided?
   c. What can we do differently next time?
   d. What will it take?

For this process to be successful and meaningful, supervisors must also critique their technicians during and after each job, as a normal and routine element of their on-the-job supervisory responsibilities. Supervisor to planner feedback and even technician to planner feedback should not necessarily have to wait for the weekly managers meeting. Focused or specific feedback should be part of ongoing team effort that occurs at the earliest opportunity and should always be provided in a constructive manner. Some operations solicit mechanic feedback by a Job Plan Survey.

While management can assess the quality of planning by periodically requiring completion of the "Job Plan Survey" the ultimate assessment of planning quality occurs at each periodic work sampling, which quantitatively determines that portion of the trade's effort actually devoted to direct productive work (on-site use of tools/wrench time) and reports improvement trends resulting from planner efforts.

Job Plan Survey User Instructions

The following is intended to instruct maintenance planning supervisors, trade and area maintenance supervisors who supervise work and other personnel who may benefit from use of the *job plan survey*. The purpose is to gain useful feedback information concerning a job that could be helpful with a particular job or future jobs.

1. Maintenance Planning Supervisors

Maintenance planning supervisors may initiate a job plan survey for selected jobs for the purpose of monitoring quality of job plans. It may be initiated prior to the work and accompany the job order. They may also implement it upon completion of jobs when it is obvious that there have been large deviations from the job plan, resulting in a savings or cost overrun.

2. Work Supervisors

Supervisors responsible for the execution of work should initiate the job plan survey as a means of formal advertisement that a job plan needs or needed improvement and/or changes. It should also be used to identify and highlight reasons for delays, cost overruns, and savings.

3. General
   a. When possible, a note should be included on the job order that a job plan survey has been issued and is to be completed.
   b. The "Job Planning Survey" form should be completed by the supervisor overseeing the work. The supervisor should seek input and assistance from appropriate hourly technicians.
   c. Completed job plan surveys should, whenever possible, be reviewed with the appropriate hourly technician, the supervisor overseeing the labor, the planner and the senior maintenance planner/supervisor.

Job Planning Survey
Your participation in the job plan survey is completely voluntary. Your honest evaluation will be appreciated. Its purpose is to do a better job in planning other jobs.

Trade_____   J/O _____   Planner _____

Work Description _____

Circle or check your response and print explanations as neatly and complete as possible.

1. Job instructions were (clear, vague, misleading, incomplete, other - explain)

2. The estimated work force was (about the right size, too small, too large, other -explain)

3. Actual work performed was (less, more, the same) as the significant work indicated to be performed on the Job Order.

---

4. How frequently would you estimate that this type of work or job is performed?

5. Were there any unusual or unexpected problems as compared to similar work or doing the job previously? Explain.

6. Were trips made, after the job had commenced, for parts, materials equipment or tools? Y{ N{ Explain.

7. Were there any problems or delays with permits or having the equipment available to work on? Y{ N{ Explain.

8. Was the work held up in any way because of other tradework, which needed to be performed first? Y{ N{Explain.

Signature _____

Supervisor _____

(Please return completed surveys with job order)

Management Assessment of Planning Quality. Basic questions for management to ask periodically about planning include those that follow.

Yes   **Question**

❏   Are all work orders filled out correctly?

❏   Are work orders analyzed in the field?

❏   Are supervisors given an opportunity to contribute to the planning and scheduling of work orders?

❏   Is the justification for work orders and particularly the lead-time allowed questioned/validated regularly?

❏   Are the predetermined materials and equipment needs specified for all work orders?

❏   Are estimates demanding but realistic? Is feedback from the supervisors encouraged? Are estimates being steadily refined?

❏   Are the least required number of technician assigned to jobs whenever possible?

❏   Are sketches and specifications made when required?

❏   Is the planning of work orders up-to-date (i.e., unplanned backlog kept at 20% or less of total)?

❏   Are recurring jobs analyzed for the purpose of establishing model work order plans?

❏   Are work orders properly coded as to type of work and is the correct authorization obtained?

❏   Are daily planning and scheduling visits held with the customers and supervisors according to the procedure?

❏   Is shutdown information obtained sufficiently far enough in advance to plan effectively?

❏   Are approaching shutdowns given attention soon enough to plan adequately?

❏   Are work orders "Closed Waiting for Sign Off" and "Waiting Material" regularly checked as to status?

❏   Are contract jobs properly charged to work orders?

❏   Is a full day's work scheduled every day for every maintenance person?

❏   Is operations notified in advance when to have equipment shutdown or prepared so that it can be worked on without delay?

❏   Are daily schedules consistently issued on time?

❏   Does the supervisor have faith in the schedule and follow it?

❏   Are work orders scheduled according to the priority established by the originator? If a job cannot be scheduled within the desired interval, is the originator notified?

❏   Do daily work schedules account for every technician clearly, including absences?

❏   Does the process of scheduling force a review of every job in progress each day?

❏   Is the backlog reviewed regularly to identify overdue work orders and action established with the custodian?

❏   Are arrangements made with the custodian to establish who will arrange for special safety or entry permits?

❏   Is shop work coordinated closely with fieldwork?

❏   Are completed work orders promptly returned to the originator to close?

❏   Are the necessary labor resources scheduled for minor repairs?

❏   Is the effectiveness of the estimates and plans checked after job completion?

❏   Is the delivery of predetermined material arranged for in advance?

❏   What effort is made to improve material specification and insure its availability on the job when needed?

❏   Are preventive maintenance inspection sheets checked and the necessary work orders written and scheduled?

❏   Do the work schedules contain a backup of lower priority unscheduled work (2 or 3 fill-in jobs)?

❏   Are all PM work orders properly scheduled according to the frequencies established?

❏   Is follow-up maintained on all orders for materials?

❏   Is the backlog of corrective work orders under control?

❏   Is there an environment of order, discipline and efficiency displayed at the planner's desk?

Functional Goals and Measures of Performance. Planner performance is also measured quantitatively, just as maintenance performance, production rate and quality and any number of other plant functions. Properly defined metrics are the ultimate performance measure as they are not swayed by personalities, biases or whimsy. The maintenance planner, the planning function, is gauged by:

- Improved plant conditions and improved use of labor as measured by a reduction of emergencies (goal is 10% of maintenance labor resources).
- Satisfaction of required job completion dates as specifically requested or implied by final priority.
- Improvement in Mean Time Between Failures (MTBF).
- Maintenance of backlogs within specified control limits (Ready between 2 and 4 weeks. Total between 4 and 6 weeks).
- Timely and accurate preparation and distribution of meaningful control reports.

The planner should be assigned 100% to the planning function, and not switched around to fill in for someone who has gone on vacation. The planner must avoid getting involved in unscheduled emergency and urgent work, if he/she is to get any planning done. The first day, from a planner's perspective, is next week.

Summary of a Planned Job. Job planning encompasses the coordination of various job inputs (material, labor, procedural direction and equipment), as required, to achieve a job output of orderly completion at least overall cost. Supervisors are relieved of much indirect activity, enabling them to spend their time more effectively by overseeing the trade crews while the planning function is effectively performed by para-managerial personnel.

Despite being the key to maintenance effectiveness, "planning" has different meanings for different people, depending on background, experience and application. The best understanding is achieved by establishing the criteria of a planned job:

1. Need is shown—a work order exists outlining scope of the work;
2. Analysis is thorough—the job has been broken down to individual components;
3. Required skills—are identified and time estimates made;
4. Material needs—are identified, ordered and on hand for timely availability;
5. Special tools—to perform the job have been gathered;

6. Required specifications and drawings—are at hand;
7. Preparatory and restart-up activities—are listed and prepared for scheduling.

When these planning steps are complete, the job can be scheduled according to priorities established with Operations and others involved the internal customers. When the work is assigned, technicians are productive because delays have been anticipated and forestalled.

6.1.6.1 Building a History

Maintenance history for specific equipment is one of the foundation elements of maintenance management. It is essential to refinement of the preventive/predictive maintenance program and is the primary tool of reliability engineering in evaluation and analysis of the current program in order to direct necessary refinements. *Equipment history* also supports the information needs of engineering, operations, accounting and other members of maintenance.

Meaningful and readily usable and retrievable equipment history is dependent upon a thorough, intelligent and consistently utilized equipment numbering system. Equipment history systems that are properly designed and effectively administered facilitate:

- Identification of equipment requiring abnormally high levels of maintenance.
- Analysis of maintenance history for the high maintenance equipment to identify specific repetitive failures to which engineering discipline should be applied to determine how equipment or instrumentation might be modified to reduce premature equipment failures, frequency of repetitive failures, and the general level of required maintenance.
- Comparison of equipment maintenance cost with replacement cost as a tool in capital planning.
- Justification and refinement of the preventive maintenance program.

Equipment maintenance history is primarily the result of data generated from completed work orders. The maintenance management information system (CMMS or EAM) should contain the capability to generate on demand the history of work order activity for any piece of equipment to which unique identification has been assigned. There is also a growing trend that includes specified production data (e.g., production capacity—dates, manufactured product lot numbers, etc.) in the Equipment History. In the event of post-consumer identified quality problems, the production data contained in Equipment Histories can be invaluable to Reliability Engineering. Retrieval of both maintenance and production history constitutes an important analytical tool by which reliability engineers analyze significant maintenance trends, predict problems that are developing and identify equipment modification or redesign criteria. Resultant corrective actions of the analytical process include improved repair procedures, replacements, modifications, upgrade and redesign and adjustments to content or frequency of the PM/PDM program.

To provide for the accumulation of equipment history, it is necessary to establish a reference (numbering) system to identify processes, equipment, components, instrumentation loops devices, etc., for which it is believed that history would be useful. The equipment numbering system is also the CMMS first-order identifier of plant equipment and therefore should be developed jointly by Maintenance and IT personnel. Although it might be desirable to maintain data on every item and component of equipment and instrumentation, the benefits to be derived are not worth the administrative effort to collect such information. To avoid such a situation, the following criteria are applicable when defining items requiring assignment of an *equipment identification number*:

- item is readily identifiable;
- item is large enough to be meaningful, yet discrete enough to permit accumulation of valid data;
- item is one where preventive or predictive maintenance requirements currently exist or are anticipated;
- item is one where maintenance cost and repair histories would be of value to engineering, maintenance, operations, accounting or others.

Assigning a number to a specific part within a component is excessive. The level of detail can be derived from work order descriptions and/or from stock usage records. However, the identification of the point where equipment-numbering stops must be clear so that items from that point are picked up within the work order description. This understanding must be conveyed to all personnel authorized to originate work order requests.

Although they are often confused, the *equipment history* and the *equipment record* differ.

Equipment History has been discussed above. It contains a database of work orders, and possibly production data, completed against specific equipment, listed by Equipment Numbers assigned in accordance with the chosen and designed CMMS numbering scheme. Some systems accumulate all work orders; others accumulate only significant work orders. In traditional manual systems, equipment history was commonly referred to as "the fat files." The fattest file was where maintenance engineers focused their attention.

Equipment Records contain nameplate, technical and original installation data, also listed by Equipment Numbers assigned in accordance with the chosen CMMS numbering scheme, for *every* piece of equipment. Equipment Records must be updated, whenever equipment is modified, replaced or completely overhauled, with details of the configuration change, cost information, dates, etc. Together with Equipment History, these two databases provide a *complete* record of the equipment Life Cycle and Life Cycle Costs.

History information on equipment maintenance is typically sought in two forms:

1. Installed location: the primary data-gathering form for maintenance information as well as cost accounting.
   a. Equipment history is primarily concerned with the magnitude and nature of repairs at a specific process point, as the atmosphere, application and usage at the installed location is normally the principal reason for high maintenance cost.
   b. To support various control systems, work order costs must be charged to a proper account to accumulate costs by that organizational unit most directly responsible for the magnitude of required maintenance.
2. Specific equipment unit: As the repair/replace decision is normally related to a specific unit, this secondary sort of equipment history is also desirable. It requires accumulation by unit regardless of where the unit may be installed. Unit accumulation should be selective, but careful design of the computerized work order system, coupled with effective equipment identification coding, can yield both sorts of equipment history as needed.

Equipment History (What and Why)

- Equipment history is a foundation element of maintenance management. It is a primary tool of reliability engineering.
- Identification of equipment requiring abnormally high levels of maintenance.
- Analysis to identify specific repetitive failures.
- Comparison of maintenance cost with replacement cost.
- Justification and refinement of the PM program.

- To evaluate maintenance failure trends in order to direct corrective action, reliability engineers need reliable, meaningful and detailed history of repairs.
- Equipment history, equipment records and equipment downtime reporting are often confused:
  - Equipment History is a maintenance-engineering tool.
  - Equipment Records are a planning tool.
- History also supports the informational needs of engineering, operations, accounting and other members of maintenance.

From the foregoing, seven essential elements for effective Equipment History can be identified:

1. An effective equipment numbering system:
   a. Installed location.
   b. Specific equipment unit.
2. A well designed and administered work order system.
3. Effective cost distribution to work orders (labor, materials and contractors).
4. Accurate downtime reporting.
5. Meaningful and consistent work descriptions.
6. Ease of information retrieval.
7. Reliability Engineering to make effective use of the information base.

6.2 CLOSING OUT WORK ORDERS

It has been said that . . .

If the plant management information system is the global economy, then the maintenance work order system is the national economy.

The WO system needs to efficiently handle huge amounts of information and many kinds of input and output, yet be understandable at its interface by the least skilled operator, clerk and maintenance worker. In other words, a good WO system has internal complexity that is transparent to its users; transparent because you have developed lists and tables of standard coding and standardized wording that describe the equipment, the failure, the repair and the resources expended. If there is one shortfall of all CMMS and EAM software, it is their inability to translate that which they do not know. If CMMS is to create and maintain meaningful history, the trade filling in the completion portion of the work order must know *and* understand the standards.

Printing all of the tables of codes, words and phrases on the reverse of the work order form has been attempted, however only one person in 10 could read them because the print was so small (to contain all of the information). Here are a couple of alternative methods.

- Utilize a Work Order Clipboard that has the coding tables laminated onto the front and back.
- Use a metal, lidded "document carrier" that has laminated copies of all the coding tables tethered inside.

There are undoubtedly a number of other ways to provide the work order standards to the trades required to complete the work order. Any one of them, or either of the suggestions above, will satisfy the "know" part of the requirement to know and understand the codes. The "understand" part of the requirement may be a little more problematic to impart. It is just human nature to be reluctant to admit, "I don't understand what that means." Therefore, it is advisable to have several short training sessions on completing work orders. Each session should include one or two of the tables and provide the definitions corresponding to each code. When covering the "verb, adjective and noun" tables, each entry should be read and a very brief descriptive provided—regardless of the simplicity or obviousness of the entry. For example, "Sheave—You know, the wheel with the groove for the drive belt." Remember, include some brief descriptive for every entry, no matter how simple, and hopefully no one will be left behind.

The work order having been "completed" with the entry of the proper codes in all of the required fields and then having the appropriate sign-offs affixed does not quite close out the work order. There is still the matter of entering all of the information into your computer information management system.

6.2.1 Data Entry and Validation

Who should enter the data? This can be an area where there is much disagreement. As the maintenance function has become more sophisticated, the number of work orders has increased dramatically. This in turn results from the focus on future analyses and reliability improvements that the modern CMMS or EAM has facilitated.

Some maintenance organizations use the planner/scheduler for this task, while others believe that allowing the technicians to do so further empowers them. Data entry does not empower, nor is it a duty that the maintenance planner/scheduler should be required to perform. The sheer amount of data entry that modern CMMS employs requires the dedicated services of a technical clerk for maintenance assigned to the Maintenance Planning/Scheduling group. The position can also be used for many updating (equipment records for example) and data entry tasks, depending on the amount of training provided.

A review process for ensuring data accuracy and integrity when closing out work orders should be defined within the Maintenance Department SOP for the Work Order System/Planning and Scheduling processes. The basic steps to be included are:

- Completed work orders, with all completion data fields filled in, are forwarded to the supervisor at completion of the day's work or at task completion.
- Supervisor reviews the data to ensure it accurately reflects the work performed, and in sufficient detail signs and forwards to the planner (or via alternate routing as determined by WO category and priority).
- Planner reviews, through the use of a standard check-off list, the WO data for the following:
  - Changes to planning figures (Estimates, durations, material and resources, additional tips and/or changes to procedures, documentation or safety information, etc.).
  - Review of coding for compliance with standards.
  - Requirement for additional work orders to cover correction of abnormalities noted by the trade workers or caused during the execution of the work.
  - Verification that WO data forwarded to Reliability Engineering (as appropriate).
- Technical Clerk enters work order completion data into CMMS/EAM. Planner reviews, using CMMS, the completed Work Order.
- Exception reports are analyzed by the planner and/or scheduler to ensure that no work order has slipped through the system and to identify changes to WO planning/estimating libraries.
- Additional elements required to meet management goals and mission.

6.2.2 Reliability Engineering

Reliability Engineering, or referred to in some plants as Maintenance Engineering, is the element of the maintenance operation that functions to:

1. Guide efforts to ensure reliability and maintainability of equipment, processes, utilities, facilities, control loops and safety/security systems.

2. Define, develop, administer and refine the preventive/predictive maintenance program.
3. Reduce and improve (optimize) maintenance work wherever feasible; assuring efficient and productive operation of plant, process and equipment; while protecting and prolonging the economic life of plant assets; all at the least (optimal) cost practical.

Serving in a staff capacity, Reliability Engineering relieves maintenance supervisors and planner/schedulers of those responsibilities that are rooted in engineering skills. Reliability Engineering is the prime user of equipment history information. Without the function, this key feature of any work order system will be ineffective, under utilized and cost prohibitive.

Reliability Engineering is different and distinct from plant engineering. Plant engineering supports management's longer-range capital program of new installations, product improvement and new product development. Reliability Engineering is dedicated to addressing production equipment reliability at the lowest cost. Reliability Engineering has two principal responsibilities:

1. Development and refinement of the preventive/predictive maintenance program in a systematic, professional manner—coordinating all efforts with equipment custodians and maintenance management to ensure the program is properly approved and endorsed by both departments.
   a. Coordinate with planner/schedulers to ensure that correct PM/PdM services and inspections are scheduled and performed when due.
   b. Provide administration, control and analysis of all data, records, and equipment histories to continually review and refine PM/PdM frequencies.
2. Development of engineering solutions to repetitive equipment failures and other maintenance problems; implementing ways of reducing the need for maintenance and ultimately eliminating the occurrence of failures. Repetitive failures are identified and the PM/PdM is refined by dynamic and relevant:
   a. review of predictive and other inspections;
   b. analysis of equipment histories, maintenance cost reports, and failure reports;
   c. review of service and inspection records;
   d. analysis of work order data.

PM/PdM benefits are realized more quickly and are better sustained when these two responsibilities are performed by reliability engineers wholly integrated as a group into the maintenance operation, rather than being individually assigned to supervisors, planners or plant engineers. Reliability Engineering duties are to:

1. Develop and administer an overall preventive/predictive maintenance program, update existing programs, add new equipment to existing programs.
2. Specify repair techniques for major repetitive tasks, such as component replacements, develop standards specifying the end product and the resources needed for these tasks.
3. Ensure responsible personnel are trained in carrying out the programs.
4. Analyze equipment histories to identify specific repetitive failures and effectively address identified areas.
5. Review developed PM/PdM procedures with operation superintendents and staff, securing "owner" approval prior to implementation.
6. Analyze planned idleness of operating equipment to develop and improve PM/PdM schedules.
7. Refine work content of PM/PdM routines to improve methods.
8. Periodically compare PM/PdM frequencies to downtime reports and equipment histories to identify where frequencies require adjustment—up or down. Refine the frequencies on an "economic" basis to avoid "over maintenance."
9. Review all equipment failures. Determine what PM action might have been taken to prevent failure in order to protect against reoccurrence. Revise PM instructors accordingly.
10. Regularly review downtime reports and equipment histories to identify reoccurring maintenance problems requiring engineering attention. Find engineering solutions to the identified situations.
11. Work with maintenance planners, as necessary, to develop job scopes for major unique jobs.
12. Routinely contribute to the planning of realistic production capacity based upon analysis of preventive maintenance time needs, repair histories, inspection reports, etc.
13. Apply value analysis to make maintenance decisions, i.e., repair/replace and repair/redesign.
14. Develop and standardize program that influences new construction and equipment purchases including materials, equipment and spare parts.
15. Ensure that accurate, up-to-date spare parts lists are available for existing equipment; specify standardization of components used in repair. Ensure spare parts lists are provided for all new installations and that components identified are consistent with established standards.

16. Identify potential for cost reduction through extended parts life, reduced labor cost, and other parts-related improvement techniques.
17. Participate in review phases of design of capital additions and changes in plant layout to ensure full maintainability of equipment, utilities and facilities.
18. Participate in approval of all new installations, including those done by contractors, to ensure their maintainability and reliability as influenced by life cycle costing.
19. Study corrosion, fatigue, wear and erosion rates throughout the plant and initiate corrective action as required.
20. Control selection and application of paints and other industrial coatings.
21. Identify, based on data and field observations, training required to improve trade skill levels or to improve repair techniques.
22. Establish training for selected maintenance personnel, by area in the techniques of:
    a. Vibration analysis
    b. Infrared detection
    c. Ultrasonic testing
23. Personally conduct any complicated and/or specialized diagnostic inspections and analytical procedures that require special training and experience.

Clearly, the Reliability Engineering function is a full-time responsibility. The Reliability Engineering function must always justify its existence on a profit-improvement basis. This requires that economic analysis be a part of every job. Periodic reports should be distributed covering subjects such as:

1. Accomplishments of the *preventive maintenance* program including:
    a. Overall schedule compliance with general plans of the preventive maintenance program.
    b. The effect of preventive maintenance expenditures on the total maintenance cost of selected items of equipment.
    c. The effect of preventive maintenance expenditures on the downtime of individual equipment items.
2. Alternate solutions to reduce the high costs associated with certain units of equipment.
3. Recommended economic studies for equipment retirement, modification, updating, etc.
4. A year-end report covering all aspects of the preventive/predictive maintenance program and outlining a specific detailed program to improve its function and to further reduce overall maintenance costs.

Reliability Engineering functional effectiveness results in:
- fewer failures;
- less downtime;
- lower material costs;
- improved equipment reliability;
- improved equipment operation;
- increased plant output;
- extended equipment life;
- fewer emergencies;
- more planned work;
- better resource utilization;
- reduced overtime;
- reduced contract expenditures;
- reduced maintenance cost.

6.3 LEAN MAINTENANCE PLANNING

Lean Operations are characterized as "eliminating activities and processes that are not value adding"—"doing more with less" and "achieving continuous improvement." A properly established, organized and trained planning function brings value to businesses in excess of their costs. More maintenance work is accomplished in less time using the same resources than would be the case if the planning function did not exist.

Once Maintenance Planners have mastered the basics of backlog management and job planning, their foremost concern should be obtaining meaningful feedback from supervisors and tradespeople regarding planned job packages. As time permits, visits to work sites can be the most effective method to achieve continuous improvement. Even though surveys, questionnaires and manager evaluations will provide meaningful input to the planner's efforts, nothing can be as effective as actually witnessing the use of his or her product. The planner will recognize elements of the job package that can benefit from revised thinking that the trades performing the work may never recognize simply because they have "always done it that way."

If the bottom line is not improved by having a planning function, it is usually the result of poorly defined roles and responsibilities, an absence of understanding of the planning role and its value, a lack of support from management, insufficient planner training, or having the wrong people in the planning role. Technical skills are extremely important in selecting and designating a maintenance planner. Many will suggest however, that the planner's personal skills are even more important. Many aspects of the planner's responsibilities involve persuading his or her immediate, or higher, level of supervision to provide and dedicate support for various efforts. Other aspects require obtaining meaningful criticism of his or her work from both peer level supervisors and lower level trades. The skills to accomplish these "feats" are far from being universally available and seldom can they be learned. When a Maintenance Manager is about to assign Maintenance Planners, he or she should keep these position requirements uppermost in mind. A review of the Position Description for Planners and Schedulers in Appendix A would also be quite helpful.

7

Performing the Maintenance Scheduling Function

7.1 SCHEDULING

The scheduling function puts the work into the hands of the Maintenance Organization's tradespersons. Once work has been planned and equipment and material availability has been assured, work scheduling can be addressed.

The scheduling of maintenance work, including associated coordination with the equipment custodian, is the process by which designated resources and resource skill levels required to complete specific jobs are allocated. The allocated resources are further coordinated and synchronized to be at the proper place at a designated time, with necessary access, so that work can be started and proceed to completion with minimal delay, within the intended time frame and in accordance with predetermined priorities and budgets.

In simpler terms, the purpose of scheduling is to ensure that resources—personnel and materials—are available at a specified time and place when the unit on which the work is to be performed will also be available. Scheduling is a joint Maintenance/Operations activity in which maintenance agrees to make the resources available at a specific time when the unit can also be made available by operations. Work should be scheduled to have the least adverse impact on the operations schedule while optimizing the use of maintenance resources, especially labor.

On the start up of any new maintenance management implementation, scheduling should be viewed as the *point* element, the advertising (i.e., most visible) arm of the program. Scheduling necessitates early, positive participation of the users of maintenance service and yields the earliest tangible (often within weeks of start up) results. By contrast, preventive maintenance, equipment history and Reliability Engineering require the investment of several months time before yielding measurable results. In the meantime, users are asking the question "What are *we* getting from all this?" Success of a newly instituted Planning and Scheduling effort demands that this question not be allowed to linger. The overall maintenance challenge is to create a Maintenance Operation that is both effectively responsive in terms of the customer and intrinsically efficient. Work scheduling is the vehicle that facilitates the ability of maintenance to meet the challenge.

7.2 ORGANIZATIONAL CONSIDERATIONS

Throughout this chapter, references to organizational levels, positions or titles, correspond to those shown on the organizational structure depicted in Appendix C, Exhibit C-1: *Standard Lean Maintenance Organization*. The purpose of providing an illustration of this structure is to enable individuals to determine the equivalent levels, positions and titles in their organization's structure, should it differ from that shown.

In Chapter 5, alternative Maintenance Organization schemes were discussed in significant detail. As this text concerns itself with not just Maintenance Planning and Scheduling but Maintenance Planning and Scheduling as performed in the Lean Maintenance and Lean Plant environment, most organizational considerations are pre-determined by that environment. A quick review reminds us that the lean maintenance operation is effectively practicing Total Productive Maintenance (TPM). In TPM, there are clear assignments of responsibility for the three basic *maintenance responses*:

1. routine (preventive);
2. emergency (breakdown);
3. backlog relief.

TPM also is organized to recognize three distinct (separate but mutually supportive) *maintenance functions*, so that each basic function receives the primary attention required:

- work execution;
- planning and scheduling;
- Maintenance Engineering.

The three principal types of *maintenance demand* are routine or preventive (*including Predictive Maintenance and Condition Monitoring*), emergency and planned work. The most common structure is composed of three major operating groups, each dedicated to one of the three principal types of demand. The basic concept of this structure is the establishment of two minimally-sized crews to meet both the routine and emergency demands and a larger third group devoted to planned maintenance work (ready backlog). In addition, there is an element defined by the combination of TPM and Lean Practices that has been designated as Empowered Equipment Management (EM) Teams. These teams, assigned under Maintenance Supervisors or Assistant Production Managers, consist of Maintenance Trades, Production Equipment Operators and Reliability Engineers. Depending on the size of the plant and the scope of individual production elements, a Reliability Engineer may be assigned to multiple teams. The aim of EM team efforts is to optimize overall equipment effectiveness, optimize safety and eliminate breakdowns through a thorough system of Equipment Management. These efforts include operator-performed autonomous maintenance, or minor routine maintenance, throughout equipment's entire life span. Therefore, under Work Execution the structure appears as shown in Figure 7-1.

Immediately you can see the potential problem—a portion of the (maintenance) work execution function is performed in two departments. Is this really a problem? It should not have to be. Even in the combined production–maintenance style of organization, the Maintenance Manager retains ultimate responsibility for all plant maintenance operations, including those performed by personnel assigned within the production department and

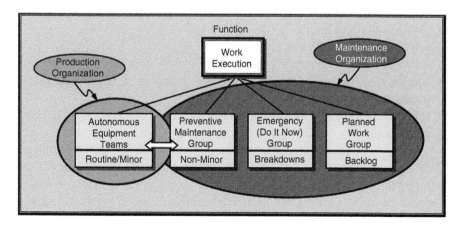

Figure 7-1 Functional Structure—Work Execution

under the supervision of a production department supervisor. In Figure 7-1, note the white arrow extending between the preventive maintenance group and the autonomous maintenance (AM) personnel; those production operators in the empowered EM Teams that perform the minor, routine and other assigned preventive/predictive (PM/PdM) maintenance work. The interpretation of the arrow is that the preventive maintenance work execution group defines the routine, preventive maintenance tasks to be performed by autonomous equipment operators and ensures that it is properly executed. Additionally, the preventive maintenance group (1) performs the remaining preventive maintenance tasks in concert with autonomous maintenance EM Teams, (2) coordinates with AM for performance of preventive maintenance and (3) augments and oversees training of AM operators (the trades assigned to the production EM teams have primary day-to-day operator maintenance training responsibility).

What does this have to do with the maintenance scheduling function? The scheduler is responsible for validating the staffing levels of the preventive maintenance group by monitoring their work efficiency and labor hour requirements for preventive maintenance tasking. The scheduler schedules those PM/PdM tasks that are due, then monitors and tracks all PM/PdM to ensure 100% completion at the designated frequencies. In addition, the scheduler ensures that the preventive maintenance group has sufficient access to equipment to perform the required preventive maintenance through coordination efforts with production managers and supervisors; lastly, he or she lists all preventive maintenance tasks on the weekly maintenance schedule and tracks schedule compliance. Summarizing the scheduler's role with respect to the Preventive (PM/PdM) Maintenance Group and autonomous EM teams, he or she:

- verifies compliance with Preventive Maintenance Program;
- validates staffing levels,
  - monitors work efficiency;
  - tracks labor hour requirements;
- coordinates/ensures adequate equipment access for PM/PdM maintenance tasks;
- schedules weekly preventive maintenance and tracks schedule compliance.

The maintenance scheduler's role in interfacing with the Emergency (Do It Now) Maintenance Group is one of monitoring. The scheduler must determine as early as possible what is the impact of emergency, breakdown/break-in work on the existing schedule. Emergency work can impact scheduled work by the following scenarios:

1. Scheduled equipment becomes unavailable.
2. A resource shortfall occurs due to diversion of those resources to augment emergency team.
3. Lack of emergency work consumes the pool of stand-by work.

The remainder (and the majority) of the scheduler's effort is spent in coordinating and scheduling ready backlog work for the planned work group.

7.3 SCHEDULING DEFINED

Having covered the scheduler's role and interface with two of the three work execution functions, Preventive and Emergency Maintenance, we come now to the third function, which is Planned Maintenance. Obviously, this is where the overwhelming majority of the scheduler's efforts will be expended.

Control in management, though much slower in reacting, is similar to the thermostat principle as illustrated in Figure 7-2. The fundamental requirement is an objective toward which progress is controlled by applying appropriate action to achieve the objective. The resultant achievement, measured against the original intention, provides feedback for correcting deviations.

The primary objective of maintenance, at the operational level, is to achieve maximum equipment reliability. The control function is in the form of maintenance schedules, which are basically statements of *when* jobs will be done. In addition, schedules represent:

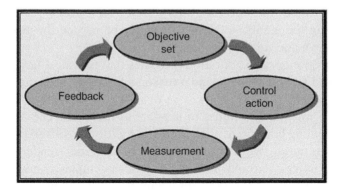

Figure 7-2 Scheduling Control Follows Thermostat Principle

1. the best utilization of personnel who can be predicted for the work that has to be done;
2. a statement of priorities mutually acceptable to maintenance and operations;
3. a means of communication for coordinating maintenance commitments between trades and with operations;
4. a definition of the maintenance supervisor's responsibilities;
5. a means of controlling time spent on each work order;
6. a working plan from which the maintenance supervisor can assign personnel, and on which they can indicate schedule interruptions;
7. a means of keeping maintenance and operations fully aware of what is happening so they can actively participate in establishing and adjusting priorities.

Scheduling, just as every operation performed in the Lean Maintenance Organization, is performed to achieve optimum equipment reliability. In effect, the schedule is the device for lining up jobs that are waiting to be done so that operational needs are best served and the best use is made of the human resources as well. The principles of scheduling are simple, but doing it is not quite so easy. Two operations must be carried out simultaneously to compile a viable schedule:

- *Resource Level Scheduling* (also referred to as *capacity* scheduling), to ensure maximum labor utilization and to avoid having persons being required in two places at the same time.
- *Job Scheduling*, to sequence the jobs on a day and time, priority basis. This is the schedule that the maintenance supervisor (and/or scheduler) uses to allocate and assign personnel and to control work.

Although these are separate operations, they must be performed concurrently. Three factors combine to make scheduling difficult:

1. predicting the number of hours that may be needed for augmenting breakdown and urgent job resources;
2. planner-developed job duration estimates that provide a target on one hand and schedules that stand a reasonable chance of being achieved on the other;
3. the need for multiple trades to support a single job. The complexity of scheduling both single trade jobs and multitrade jobs so that schedule times mesh, without idle time, for all trades render schedule creation at least ten-fold more difficult. Such complexity magnifies the desirability, almost necessity, for acquiring or developing multiskilled maintenance labor resources.

When laying out a schedule, the key element is duration of the job. This is how long the job is going to take—a very important consideration when the shutdown of operational units as well as the coordination of other trades may be involved. Quite clearly, with three difficult-to-predict and complex to reconcile variables involved, the schedule cannot be an absolute, rigid plan. Nonetheless, it must represent the most desirable objective for the maintenance supervisor to achieve. Additionally, within this framework, the maintenance supervisor must have adequate flexibility with respect to the sequence in which he or she does specific jobs. However, under no circumstances can the maintenance supervisor be allowed to substitute other jobs for those on the schedule without the authority of his or her immediate supervisor and the knowledge of the scheduler. The supervisor's immediate supervisor, in turn, will have to seek guidance when difficult or conflicting priorities arise.

7.3.1 Prerequisites for Effective Scheduling

Achieving scheduling objectives requires adherence to several proven principles and procedures:

1. Lead time—needed work must be identified as far in advance as possible so backlog of work is known and jobs can be effectively planned prior to scheduling.
2. Backlogs must be kept within a reasonable range. When planning and scheduling are performed by separately assigned persons, considerable dialog regarding backlog control must be maintained between them. Backlog below minimum does not provide a sufficient volume of work to accommodate smooth scheduling. Backlog above maximum turns so slowly that it is impossible to meet customer needs on a timely basis or requires an inordinately large amount of overtime.
3. Special or heavy demands cannot be scheduled unless backlog is addressed by providing additional resources or by relaxing/reassigning priorities.
4. Jobs will not be scheduled until *all* planned needs (parts, materials, tools, special equipment, the item to be worked, any special support) are available in the quantity required and at the time necessary.
5. Each available maintenance trade must be scheduled for a full day of productive work for every day of availability.
6. Emergency work may be done at the expense of scheduled jobs if additional resources are required to augment the Emergency Work Group. The displaced scheduled jobs would constitute an overloaded schedule

and result in work being carried over to the next schedule period unless addressed by a temporary increase in capacity, i.e., overtime.

7. Additional work (amounting to approximately 10% of available scheduled labor resources) will be identified and posted to the schedule as fill-in work for situations where scheduled jobs cannot be performed for a legitimate reason or other scheduled jobs have been completed in less time than planned.

Adherence to these prerequisites ensures that:

- all maintenance needs are properly attended to;
- accurate evaluations are made as to the importance of each job with respect to the operation as a whole;
- customers have their work performed on a timely basis;
- equipment downtimes experience minimum delay;
- work is performed safely;
- overall maintenance cost is kept to a minimum.

All these combine to reduce the overall cost of a quality product.

"Lack of forethought on your part,
does not constitute an emergency on my part!"

7.3.2 Preparing Schedules

It is often more convenient and practical to prepare the weekly maintenance schedules using a daily scheduling approach. Daily schedules can then be merged into weekly schedules and used for time distribution purposes as well. The basic steps for preparing and using schedules are as follows:

1. The Scheduler prepares a schedule form for each supervised maintenance unit by filling in week beginning date, name of the responsible Maintenance Supervisor and the unit involved.
2. The Scheduler determines the amount of work that needs to be scheduled based on the available crew capacity and the trend of Ready Backlog size. (i.e., Are ready backlog labor hours within the 2–4-week range and has the ready backlog been steadily increasing, steadily decreasing or varying within a reasonable range inside the 2–4 week level?)
3. The Scheduler determines, by reference to weekly resource reports and, if necessary, by discussion with the Maintenance Supervisor, the

quantity of labor resources (labor hours, by trade) expected to be working during the schedule week.

4. The Scheduler and the Maintenance Supervisor compare the amount of work needed to be scheduled to available resources to determine whether overtime (or some other form of labor resource level adjustment) needs to be arranged for.
5. The Scheduler, in concert with the Maintenance Supervisor, if desired or designated, reviews all work orders "Ready to Schedule" and from his own knowledge as well as from discussions with the appropriate operations supervisor, puts work orders into sequence order by priority.
6. The Scheduler lists each work order on the schedule form, which, when completed, is provided to the applicable Operations Supervisor, Maintenance Supervisor, and Support Activity Supervisor when applicable.
7. The Scheduler provides the updated Ready Backlog report to Operations. Operations should be trained on how to retrieve the report from the CMMS, but may require a printout of the report to be provided. Operations will review the report and select work they desire to be scheduled, or consider for scheduling, during the Weekly Schedule Coordination Meeting.
8. The Scheduler conducts the Maintenance Weekly Schedule Coordination Meeting attended by the applicable Operations Supervisor, Maintenance Supervisor(s), Preventive Maintenance Supervisor, and any other designated personnel. The meeting's purpose is coordination of maintenance and operations requirements in order to define and publish the maintenance weekly schedule.

However well schedules are constructed and coordinated, interruptions will occur and, after obtaining required authorization, the Maintenance Supervisor writes in the description and duration of the job(s) that interrupted the schedule. By using this process, the Maintenance Supervisor can keep a check on the priorities and on what is happening in the plant. By the same token, operations management and supervision can be kept aware and provide guidance on the full implication of conflicting priorities.

7.3.2.1 Scheduling Practices

Although the schedule prescribes the framework of the work for each week, it must be updated and reviewed daily by the Operations Supervisor, the Maintenance Supervisor and the Scheduler so that appropriate adjustments can be made to compensate for interruptions. The results of these daily reviews and adjustments are published daily in the form of a Daily Maintenance Work Schedule. The following guidelines are recommended as sound scheduling practices:

1. List jobs on the schedule in descending order of priority and ensure that about 20% of labor-hours are scheduled on very low-priority jobs.
2. Schedule work for all available labor-hours and, if necessary for backlog management, for all required overtime labor-hours. Use the personnel assigned to low-priority work for Emergency/Breakdown Group Augmentation, if required.
3. Involve operations management in preparation of the schedule and provide them with copies to ensure that the commitments are acceptable and understood by both maintenance and production.
4. The planner should have ensured that work order operation steps or sub work orders are written for all trades on multi-trade jobs and the scheduler verifies that and they are then scheduled at the appropriate times on the schedules for those trades.
5. Operations management must advise maintenance management at the earliest possible moment if they are unable to release equipment as scheduled. Similarly, the maintenance department must advise operations management if the reverse situation is likely to occur. Maintenance Schedulers ensure that necessary coordination is carried out by establishing an ongoing dialog with operations and maintenance management.
6. Review schedule at the end of each day and the first thing each morning to update and adjust it as necessary. The Scheduler verifies work and work order completion and that completed work orders are forwarded to the Planner (via CMMS as applicable).
7. The Scheduler ensures that "interrupt jobs" are written onto the schedule, or personally enters them on the schedule (action to be defined by SOP). The Scheduler then ensures that "interrupt job" coordination with operations and other trades is carried out, or personally coordinates with operations and other trades (action to be defined by SOP).
8. The Scheduler will rigidly follow and enforce rule that a schedule must be prepared by Friday of each workweek for each Supervisor, showing how he or she will utilize their personnel during the following week.
9. The Scheduler will ensure, via the Maintenance Manager, rigid enforcement of the rule that Maintenance Supervisors will not arbitrarily cancel jobs off their schedule.
10. Schedulers fill out schedule compliance report each week.

11. The Maintenance Manger reviews schedule compliance with maintenance unit supervisors each week. Operations and maintenance formulate plans as necessary regarding achieving/improving schedule compliance.

7.3.2.2 Scope of Maintenance Scheduling

Scheduler's Role

The scope of Maintenance Scheduling encompasses the allocation and coordination of the resources required for specific jobs. In addition, it includes the determination of when jobs get done and which resources can best be applied to their performance. The Maintenance Scheduler's role involves the following:

1. *Operations and Maintenance (Work Execution) Liaison* (for nonemergency or urgent work). Ensure that current and well-organized relevant backlog reports have been pre-issued to operations and maintenance supervision in preparation for weekly schedule coordination.

    View liaison with a particular "customer" as a permanent relationship. Learn and take interest in their problems. Remain abreast of their workloads, short- and long-term plans and priorities. Help them think far enough in advance to facilitate effective planning. Provide continuity for their maintenance knowledge, records and information.
2. *Determination of Resource Availability*. Ensure that expectations for backlog relief are realistic. Early notification of a growing backlog trend can facilitate early preparation for additional resources, e.g., overtime.
3. *Moderate Weekly Schedule Planning Meeting*. Achieve a consensus between equipment custodians and maintenance trades supervisors with regard to the most effective near-term deployment of available maintenance resources. Be a moderator as long as possible; shift role to chairperson (i.e., coercion) of the meeting as a last resort.
4. *Preventive Maintenance Scheduling*. Ensure that all preventive/predictive routines are scheduled at their predetermined frequencies. Utilize CMMS automated printout.
5. *Backlog Management*. Ensure that requested completion dates (real or implied by assigned priority) are met while, at the same time, ensuring that even low-priority jobs reach the schedule in a reasonable period of time. Ensure Ready Backlog is maintained within the range of 2 to 4 weeks of labor. (*See Appendix C for a short tutorial on the use of control charts for backlog management*.)

6. *Daily Scheduling and Schedule Adjustment*. Coordinate new, high-priority work orders with those already in the weekly plan. Always strive to optimize schedule compliance, despite essential schedule "breakers." Practice capacity scheduling by always selecting a little more than enough work for each person, each day.

7. *Support of Job Execution*. Verify that the responsible supervisor receives and understands the planning package for each scheduled job. Provide follow-up coordination to assure that all agreed-upon supportive actions of others are performed on schedule. (When planning and scheduling functions are performed by separate personnel this may be a planner function, although the scheduler should verify these actions.)

8. *Job Assignment to Specific Technician*. This is a local procedural matter. The supervisor retains this responsibility to enable the scheduler to focus more on future concerns rather than being caught up in present concerns.

9. *Schedule Follow-up*. Determine the level of schedule compliance and reasons for completion shortfalls. This is a "constructive" responsibility towards future improvement.

Scheduler's Activities

During job scheduling, the Scheduler coordinates with Operations by conducting a structured, regularly scheduled, weekly meeting intended to reach accord regarding the most important jobs to be scheduled during the approaching schedule week (Friday through Thursday). Jobs are selected from the "ready backlog" listing; those in the backlog that are ready to be scheduled (no holds for material, etc.). The recommended attendees of the weekly schedule planning meeting, together with the purpose, or intent of their attendance, are provided in Table 7-1. Preparation activities for the weekly meeting require that the Scheduler:

- has the Planner's current computerized backlog file of work orders available and waiting to be scheduled. The unavailable (unplanned) backlog requiring engineering, awaiting materials, awaiting equipment access, etc., are filed and categorized accordingly;
  - once available (ready for scheduling) they are filed by required start date – sorted by CMMS;
  - ensures Backlog Status Report is issued (by Planner) to all attendees on day preceding meeting (see Appendix C, Exhibit C-4);
  - links multiple jobs on the same equipment or in the same proximity;
- is conscious of PM/PdMs due so these can also be reflected in discussions and resultant schedule;

Table 7-1
Attendees at Weekly Schedule Planning Meeting

Operations/Maintenance Weekly Planning Meeting	
Attendees	**Purpose/Objective**
Scheduler	Moderate/Chair Meeting, Complete Weekly Schedule, Coordinate All Activities
Planner	Provide Backlog Status – Unavailable Work, Describe Unusual Planned Work Packages, etc.
Preventive Maint. Group Supervisor	Obtain, compare and review scheduled preventive maintenance printout (CMMS)
Planned Maint. Group Zone[*] Supervisor	Advise & Consent to Schedule All Resources, Identify Windows of Other Commitments
Area Storeroom/Parts Supervisor	Identify Scheduled Parts and Materials Staging Requirements
Maintenance Engineering Liaison	Attendance based on projects or tests to be planned or scheduled
Production Line/Zone Supervisor	Advise & Consent to Schedule Off-line Windows for Planned Work
Empowered EM Team Leader	Coordinate Equipment Off-line periods, Identify Autonomous Maintenance Requirements
Support Activities Supervisor(s) -by invitation	Mobile Equipment, Crane, Rigging, Fabrication, etc., Support Requirements Coordination
Safety/Environmental Representative(s)	Ensure Compliance With Applicable Codes and Regulations

[*] *When maintenance is organized differently than shown in Appendix B – for example, planned work execution is organized by specific trade skills – then each trade supervisor should attend the weekly maintenance schedule planning group meeting.*

- ○ considers approaching PM/PdMs: perhaps they should be performed early to take advantage of the scheduled downtime and avoid another shutdown in only a matter of a few weeks;
- ○ ensures the PM/PdMs of all groups are considered;
- negotiates for downtime windows, when selected maintenance can be performed, and obtains specific agreement on timing of necessary equipment access;
- ○ coordinates with Operations Planning;
- ○ makes reliable outage duration estimates;
- ○ searches for the best time to take the equipment down and perform the necessary work by considering it from all perspectives, but remains aware that the internal customer must ultimately prevail;
- ○ arranges for necessary safety inspections, fire watches and standby positions associated with ladder safety, vessel entry, etc.

When selecting jobs for the Weekly Master Schedule, all parties should be aware of those jobs approaching their requested completion dates. Any of these jobs that cannot be scheduled to meet those dates can be discussed with the WO originator in the context of the priorities of other available work as established by all attendees. In this manner, WO priorities can be "fine-tuned." Based upon input from, and agreements reached during the coordination meeting, the Scheduler prepares the Weekly Master Schedule (refer to Figure 7-3).

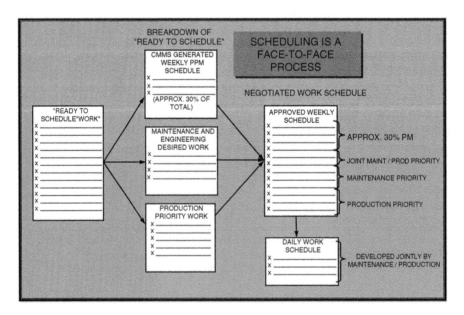

Figure 7-3 Maintenance Schedule Planning Meeting

For each job placed on the schedule the Maintenance Scheduler must:

1. verify availability of parts, materials and special tools required for execution;
2. in concert with Maintenance Supervisors, allocate available resources to specific jobs,
   a. balance work scheduled with labor-hours available;
      i. make a conservative provision for urgent schedule breaks;
      ii. schedule 100% of remaining labor-hours to the "fine-tuned" prioritized work list;
      iii. identify fill-in jobs to be performed should scheduled jobs be unavoidably delayed;
   b. prepare weekly/daily schedules in accord with established importance; this may be up to 10% of the available labor hours;
3. issue final schedule,
   a. incorporate all preventive/predictive maintenance inspections at their predetermined frequencies;
   b. schedule the timely completion of all identified corrective maintenance;
   c. review the schedule and planned job packages with the Maintenance Supervisor to ensure that nothing falls through the cracks due to misinterpretation of intent or meaning.

The Weekly Master Schedule should now be a document that all parties, through contribution, accept ownership. Requested schedule breaks require the sanction of the Operations Manager, Operations General Foreman or Shift Supervisor on the off-shifts. Appendix C, Exhibit C-5, Weekly Master Schedule, illustrates a typical layout of a weekly schedule form. The actual format of the schedule template is dependent on your particular plant's organizational structure, trade categories, skill levels and so on.

Daily Schedule Adjustment (See Table 7-2 for a typical daily schedule.)

When making adjustments to the daily schedule, the Maintenance Supervisor:

1. checks preparedness for each day of the Weekly Master Schedule by
   a. verifying that he/she has the planned job packages. The supervisor's packets should consist of not less than the following:
      i. turnaround schedule, if applicable;
      ii. copies of all jobs that he or she will be supervising for that day;
      iii. progress report forms—completed to the point applicable;
      iv. telephone and/or pager list;

Table 7-2
Typical Daily Work Schedule Entries

DAILY WORK SCHEDULE				
DAY	**Thursday**	**CREW/CRAFT**		
DATE	**10/12/2005**	**Zone 2 PM Group**		
Work Order No.	**Equip. No.**	**Description**	**Craftsman**	**Time**
04-55743-01	71021	Rebuild Bal.	Wayne	7:30AM
04-54371-01	73012	Fabricate Table	Joe	10:00AM
04-54365-02	73010	Replace Drive/ Align	Jason, Rob	12:30PM
Schedule Breakers				
04-55774-01	*72002*	*Replace/Adj. Clutch*	*Bill*	*3:00PM*

> v. the work schedule for supervisory personnel;
> vi. a copy of the turnaround organization;
> vii. instructions for progress reports and meetings;
> viii. a set of trade work rules;
> ix. instructions regarding materials handling;
> x. instructions regarding the use of contract Supervisor and clerks;
> xi. instructions regarding the reporting of contractor daily time reporting;
> xii. a list of helpful reminders;
> xiii. turnaround objectives;
> b. checks that each crew is aware of their assignments;
> c. checks with Operations supervision to determine that the equipment be available when the scheduled crew arrives at the job site;
> d. checks whether kitted materials will be delivered to the job site ahead of crew arrival;

2. verifies that applicable operations supervisors have finalized the daily schedule;
3. keeps all parties aware of schedule status and manpower availability on a daily basis, or more often;

4. coordinates schedule adjustments for the balance of the week (beyond the current day) with Operations,
   a. carries extended or delayed jobs over to next day;
   b. makes adjustments to protect as much of the original schedule as possible.

Job Close Out and Follow-Up

Prior to completing each day's job close outs, the Maintenance Supervisor reviews schedule compliance with the scheduler. At the end of each work-week, they perform a summary review. Following the reviews, the Scheduler:

- annotates schedule status of individual jobs;
- receives from the individual maintenance group supervisors, the completed work orders for those jobs which have been completed;
- verifies that all essential feedback entries have been made and forwards the work orders to the *Maintenance Planner* for standardized entry validation, backlog update and entry of CMMS completion data. *During WO closeout, the Planner* must:
  - verify proper account designations;
  - verify WO coding, but avoid inappropriate changes; checks with maintenance supervisor first;
  - review actual labor and material usage relative to estimated usage, in order to ascertain necessary refinements of the planned job package before filing for future use;
  - check proper disposition of any leftover materials;
    -equipment;
    -purchased materials and parts;
    -special tools (Maintenance Supervisor);
    -stock items;
    -free Bin Stock (Tradespersons);
- calculate Schedule Compliance and prepare associated reports (or enter data into CMMS).

7.4 METRICS—SCHEDULE COMPLIANCE AND LABOR EFFECTIVENESS

By approving the schedule, maintenance has agreed to perform the work contained on the schedule and operations has agreed to make the equipment available so the work can be performed and not demand diversion of the scheduled resources except for a legitimate emergency.

In order to determine the quality of the scheduling function and of the PM Group schedule performance, schedule compliance must be calculated and reasons for any noncompliance determined. Questions to be answered include:

- How did we do on last week's schedule?
- How many of the scheduled jobs were actually completed?
- Did maintenance not get to them or did operations deny access?
- How many actual labor-hours were used vs. scheduled labor-hours?
- How many unscheduled jobs broke into the schedule? Why were they necessary?

Answers to these questions will highlight the underlying reasons for poor schedule compliance. Maintenance personnel are fond of saying, "A given maintenance job never goes the same way twice." While this statement is basically true, they often add, ". . . therefore, maintenance cannot be measured." This part of their statement is false. The first part of the statement simply influences the precision with which you can estimate, the jobs where maintenance estimates should be applied and the measurement period required to level out the fluctuations in individual jobs.

Industrial engineers attempt to establish production standards with an accuracy of ± 5%. Considering the difference in consistency between production operations and maintenance work, expectations of maintenance estimates cannot exceed an accuracy of much better than ±15%.

If the estimates are used for determining Percent Performance (Estimated Hours÷Actual Hours) as opposed to Schedule Compliance (Hours of Scheduled Work Completed ÷ Hours of Scheduled Work), the calculation should be made weekly and should be calculated at the Maintenance Supervisor level, i.e., all crews responsible to a given supervisor. The one-week period puts several jobs into the calculation, allowing for averaging of unusually difficult jobs with unusually easy ones.

Accuracy on individual jobs is not reliable, but accuracy over the several jobs completed by a crew in a given week is acceptably accurate, particularly for measuring performance trends. Individual (as opposed to crew) performance can be calculated periodically, but only to guide necessary training to meet the critical needs and never for disciplinary reasons. The use of work measurement for discipline detracts from applications that are more important, namely backlog control, work scheduling improvements and group performance trends.

Labor Efficiency is the traditional efficiency report calculation (standard or estimated labor hours divided by actual labor hours). The measure indicates how well the crew is performing in relation to the expectation. Variance from estimate might indicate poor crew performance, inaccurate estimate, or inaccurate work scope. The labor efficiency number as a percentage is:

(Total Estimated Hours for Work Orders Completed/Total Actual Labor Hours for the Same Work Orders) × 100.

Schedule compliance is merely a number that indicates a quantitative value of work completed vs. work scheduled. While important, a more qualitative report that considers the "whys" of nonconformance is more important to more people. The schedule compliance number as a percentage is:

(Hours of Scheduled Work Completed ÷ Hours of Scheduled Work) × 100

In order to provide the most accurate compliance number, the hours of scheduled work used in the calculation should be the sum of the adjusted daily scheduled hours for each workweek. Table 7-3 provides a simple layout of a weekly compliance worksheet.

Schedule Effectiveness and Schedule Performance are two additional scheduling metrics that assist Maintenance Management to fully understand how their labor resources are being utilized. The data for these metrics comes from the same sources as Schedule Compliance and Labor Efficiency.

The schedule effectiveness number as a percentage is:

(Direct Scheduled Hours Completed/Labor Hours Available) × 100
The target percentage is 65%

Table 7-3
Schedule Compliance Worksheet

Maintenance Schedule Compliance			
Week of: _____			
Adjusted Daily Scheduled	**Hours**	**Work Completed**	**Schedule Compliance**
Monday	48	40	83.3%
Tuesday	42	40	95.2%
Wednesday	46	38	82.6%
Thursday	48	44	91.7%
Friday	44	42	95.5%
Week:	228	204	89.5%
Notes: Wednesday: Lost Rob at 12:00 to Emergency job in FAB.			

The schedule performance number as a percentage is:

(Scheduled Hours Complete (Direct & Indirect)/
Labor Hours Available) × 100   The target percentage is 80%

Before further discussion, a little background is needed with regard to the "Labor Paid Time." Table 7-4 begins with the total paid labor, which is the typically used number of 2080 hours and that is the "Labor Paid Time" value used in the calculation of Labor Utilization. The available hours at the bottom of the table (1707 hours)should be used, you say. Why? If the tradesperson is paid for 2080 hours, then 2080 is the number that should be used. Admittedly, that limits the absolutely best utilization to no better than 82%, as shown at the very bottom of the table.

Additionally, calculating utilization in this manner (and widely publicizing the methodology shown in Table 7-4) indelibly etches in the minds of each tradesperson, the often unrealized fact that the company is absorbing nearly 20% of every person's salary (not to mention things like taxes and other benefits) with no hope of gaining anything in return. The effect can only be to increase the general level of enthusiasm for improving labor efficiency.

In order to complete the labor utilization calculation, the productive, often referred to as wrench, time must be known. This portion of labor utilization relates to measuring how effective the planning and scheduling function is being performed. Effective, and continuously improving, planning and scheduling in a Lean Maintenance Environment is critical to improving

Table 7-4
Determining Available Labor Hours

Labor Hours Available vs. Labor Hours Paid		
Paid Labor Hours	**52 weeks × 5 days × 8 hours**	**2080 hours**
Less Adjustments		
Vacation (average)	12 days × 8 hours	96 hours
Holidays (typical)	10 days × 8 hours	80 hours
Sick (average)	4 days × 8 hours	32 hours
Break Time (typical)*	2/day @ 15 minutes each	117 hours
Training, Welfare, etc.	6 days × 8 hours	48 hours
Available		1707 hours
Available Percentage	**1707 hours ÷ 2080 hours**	**82%**
*260 days–26 days = 234 days; 234 days × 0.5 hours = 117 hours		

labor utilization. Labor Productive Time does not include time consumed by the following:

- waiting on parts or locating parts/parts information;
- waiting on other asset information such as procedures, drawings, technical manuals, etc.;
- waiting for the equipment to be shut down;
- waiting for other trades to complete their portion of the work;
- any other delays due to the lack of effective planning and scheduling;
- any other delays due to "other factors."

Accurately defining wrench time is completely dependent on accurate and detailed completion (recording job performance information)of the work order. The use of CMMS may actually detract from the accuracy of the details unless a PDA or similar information recording device is utilized. Identifying work delays and their causes after the job is completed is dependent on recall from one's memory, which, after the work is completed, might not be as accurate as "real time" identification. It is strongly recommended that the lead tradesperson on every work order be provided with a standard form for the real time entry of important work execution information. A suggested format for this information entry is shown in Chapter 6, Section 6.1.6.

One key part of planning is determining the scope of the repair job and the special tools and equipment that are required for a quality repair. A continuing concern of the maintenance planning function should be on improving existing repair methods whether by using better tools, improved repair procedures or diagnostic equipment and using the right skills for the job. Providing the best possible tools, special equipment, shop areas, repair procedures and craft skills can be a key contributor to continuously improving work plans. The Job Plan Survey process is one method to capture this information.

Another powerful metric of tracking performance of the scheduling process captures the percent of work that was planned for the scheduled week. Trending this metric on a weekly basis provides a clear picture of the relationship of labor planned versus breakdowns.

The percent planned work number as a percentage is:

$$(\text{Hours on Planned Work/Total Hour Worked}) \times 100$$

7.5 GENERAL SCHEDULING CONSIDERATIONS

Communication is the key to successful maintenance scheduling; this involves everyone from the Planner, Scheduler, Maintenance Supervisor,

Craftsman, Storeroom personnel, Production Manager and Supervisors, to the Operator who is responsible to have the equipment secure and ready for maintenance. Any breakdown in this communication diminishes the probability of success.

In the case of the plant just installing or initiating the Maintenance Planning/Scheduling function, do not utilize an automated scheduling function. Trying to automate something that is not yet been performed in the basic, manual mode will cause frustration and a general lack of understanding the process. The potential benefits of scheduling automation are best achieved by first establishing sound communication foundations that support the "by-hand" scheduling processes. By sticking to the basics, most organizations can achieve significant skill levels in their maintenance scheduling capabilities. Even when the scheduling capability exists in an already implemented CMMS/EAM, resist its use until your schedulers have mastered the fundamentals.

When scheduling longer duration and/or generally more complex jobs, the Scheduler should be alert to identify areas of risk. Look for those job elements that, if not prepared, ready, successfully executed or otherwise confounded, can seriously impact job duration, equipment downtime or craft availability for subsequent work. When risk areas are identified, work with the Planner and Maintenance Supervisor to identify potential risk mitigation actions. At the very least, visit the work site just prior to the potential risk to follow it through, being prepared to take any necessary actions such as temporarily diverting other resources, alerting production supervisor to alter production schedules or other actions as necessary.

8

Special Case: Maintenance Planning and Scheduling for Maintenance Outages—The Plant Shutdown

The majority of preventive and planned maintenance work is performed while the manufacturing plant is in operation. *Major* maintenance work, or work that, because of its scope, cannot be performed while the plant is operating, will be required at some point. Entire production lines (equipment systems) will need to be shut down for major equipment overhaul or even replacement. This shutdown is referred to as a *Maintenance Outage*. While it may be possible to just shut down that portion of the plant needing attention, the work is normally too disruptive to continue operating. Additionally, labor assets to perform the daily maintenance work in the operating portion of the plant will be in short supply. Thus, the maintenance outage most often involves a total plant shutdown. Economically, the total plant shutdown also makes the most sense. It is far less expensive to simultaneously shutdown all plant operations to perform major maintenance work on all plant equipment needing it than it is to conduct more frequent shutdowns in separate areas of the plant. This is referred to as a Plant Maintenance Shutdown or *Plant Shutdown*. One last bit of terminology is that which is applied to the process of performing the major maintenance, equipment upgrade action and/or the addition of new or expanded production capabilities. This actual execution and completion of the outage work is often referred to as a Plant Turnaround or just *Turnaround*.

Plant shutdowns to perform major maintenance work are the most expensive of all maintenance projects not only because of the loss of production, but also due to the expense of the major maintenance being performed. Industry surveys report that between 35 and 52% of maintenance budgets are expended in individual area or whole plant shutdowns. These figures reflect only the price of maintenance and do not encompass the lost opportunity costs of no production.

Shutdowns are a time of accelerated activity, with numerous vendors, contractors, and heavy equipment engaged in multiple tasks in close quarters. From 1995 through 2000, Occupational Safety and Health Administration (OSHA) records show that more than 25% of lost time accidents for any given year in manufacturing plants occur during major maintenance outages. Plant shutdowns can be complex, not only due to the nature of the work to be performed, but also because of the pressure to try to force as much work as possible into as short a shutdown period as possible. As the volume of work increases, the complexity of the maintenance outage increases, rendering the shutdown even more costly and, perhaps even more important, exceedingly more difficult to manage.

A plant shutdown always has a negative financial impact. This negative impact is due to the combined effect of the loss of production (sales) revenue together with the additional expenses associated with the major maintenance work. There is an overall positive return that is not always obvious to everyone, especially in the heat of battle and the steadily increasing pressure to get the work completed in the shortest time possible. The positive impacts are an increase in equipment asset reliability, continued production integrity, and a reduction in the risk of unplanned failure and resulting unplanned outages.

A major maintenance outage (total plant shutdown) is generally short in duration and high in intensity. It can cost (combined maintenance and lost production costs) as much as an entire year's maintenance budget in four to five weeks. Because the maintenance outage is the major contributor to plant downtime and maintenance costs, proper shutdown management is critical to minimizing the impact on the bottom line. *"Failure to plan is planning to fail."*

No single strategy is more important, or more often neglected or overlooked, than planning. Planning for and managing a maintenance outage in the manufacturing plant environment are difficult and demanding operations. If not properly planned, managed and controlled, companies run the risk of serious budget overrun and costly schedule delays. The Planning and Scheduling operations are central to completing an outage within budget and on schedule. Early identification of, or even the potential of, a problem relating to any element of the outage schedule, is the key to success and it is in the hands of planning and scheduling. Beginning the outage with a viable schedule, complete work packages and both material and personnel resources arranged for and available for contracted work as well as in-house efforts are the primary of all prerequisites necessary for success in executing the plant shutdown for major maintenance.

8.1 PLANNED OUTAGES DEFINED

The management and control of a planned maintenance outage, a plant shutdown and turnaround, can be broken down into five phases of activity. They are:

I. Definition
II. Planning
III. Scheduling
IV. Execution
V. Debrief and Lessons Learned

8.1.1 Phase I: Definition

During the Definition Phase, plant management must *determine* and then fully *define* the objectives of the plant shutdown. Critical issues that need to be addressed include:

- development of a Plant Shutdown Vision and Shutdown Objectives;
- start and duration of the plant shutdown;
- who will manage the turnaround;
- what equipment is to be involved;
- is equipment to be refurbished, completely torn down and rebuilt, or replaced?
- is new equipment, providing expanded capability/capacity, to be installed and, if so
  - has Reliability Engineering performed maintainability/reliability analyses?
  - has new equipment/vendor been identified and procurement actions initiated?
- what work will be contracted for and what work performed in-house?
- has the work and the objectives been prioritized?
- have time and cost constraints been established, i.e., work cutoffs?

When the Outage Manager has been designated, he should immediately take steps to appoint an Outage Scheduling Coordinator. In smaller plants, the outage manager and schedule coordinator will very likely be the same person. In larger plants, and during broadly scoped shutdowns in smaller plants, the Outage Manager should meet and negotiate with the Maintenance Manager to designate a separate Outage Scheduling Coordinator and an Outage

Committee consisting of *at least* one supervisory level maintenance person, one Reliability Engineering engineer and one senior maintenance, or MRO, storeroom person.

Planning, or pre-planning, for a maintenance outage should start as soon as the current outage is completed. Too often, planning is not begun until two or three months prior to commencing work and, more often than not, involves only the writing of job specifications or procuring material and parts. Much more needs to be accomplished during both the definition phase and the second or planning phase. The first efforts should be directed toward developing a Plan of Action and Milestones (POA&M) for the pre-shutdown period—planning for the shutdown planning effort. The pre-shutdown POA&M is a Gantt Chart (refer to Appendix D for a brief Gantt Chart tutorial) schedule of major events and the time frames when specific actions or evolutions must be achieved or completed prior to starting work. A "Rolling Wave" approach to pre-shutdown management is necessary. This approach involves the definition of greater and greater detail as the POA&M milestone events are completed. A partial POA&M is illustrated in Figure 8-1. Milestones that should be identified, for example, should include:

Time Frame (days prior to start)	*Activity/Event*
200–150	Review last shutdown debriefing report and non-scheduled shutdown summary reports for the past 12 months
150–100	All work requests due
120–90	Preliminary work lists completed
90–80	Preliminary job review meetings
90–80	Assignment of Planner / Work Package pairings

Task	Dec	Jan	Feb	Mar	Apr	May	Jun	Jul	Aug
Next Outage Comences:									OUTAGE
Outage Manager Assigned		◆							
Review Previous Shutdown Report									
Generate Risk Factor Listing									
Last Date for new Work Requests					4 wks.				
Preliminary Work List									
Priliminary job Reviews									
Planner / Work Package Assignments					2 wks.				
Engineering Design Work									
Etc.									

Figure 8-1 POA&M for Outage Planning Period

85–75	Engineering design work completed
80–70	Completion of planned work estimates
75–70	Management review of (final) planned work estimates/Final work package definition
70–55	Completion of planned work packages
65–50	Shutdown parts, material, equipment ordered/contracts placed
60–50	Master job package completed and published
55–45	Master job package review meetings
50–40	Job sequence and assignment schedules completed
45–35	Critical path network schedule completed
45	Initial area bar charts developed
45	Manpower requirements lists and leveling charts completed
40	Delivery dates for major equipment procurements established
30	Final schedule completed and issued
25	Schedule review meetings
15	Shutdown materials received, marked and locked up
14–10	Outage coordination meeting(s)
10	Shutdown management booklets published
7	Daily crew assignment board set up
5	Preliminary work begins (tags prepared, routing marked, etc.)
2	Initial staging of parts completed
0	Plant shutdown

The shutdown must reflect the business goals of the organization. The vision is the ultimate goal towards which the outage manager orchestrates the shutdown plan. Within the shutdown plan, objectives and expectations must be established early for the entire operation. Objectives should be concise and measurable as well as applicable to each phase of the shutdown and reflect the outcome established by the vision. Some typical objectives include:

- limit new or growth work to less than 5% of total shutdown work;
- 65% (or more) of shutdown work (less new capital project work) will be determined by inspection and condition monitoring versus historical data;
- zero safety incidents by contractors or plant work force, etc.

8.1.2 Phase II: Planning

The planning phase of a planned outage involves many of the activities seen in planning during normal plant operations, just on a different scale. Additionally, the planner will be involved in several activities that are not necessarily part of his or her normal routine. In-house maintenance projects must be planned for with complete work packages, including major material items such as replacement equipment, equipment modification designs and material, complete overhaul kits, equipment/system interface work (i.e., piping, electrical service, etc.). Additionally, outside support services for in-house work are often required, including items such as vendor technical reps, transportation, handling and rigging services beyond in-house capabilities, etc.

Delivery of required material to the job site should be planned just prior (JIT) to work commencing. This eliminates transit time and loitering by the work force and increases time on tools for each trade/craftsman associated with that particular job. Additional planning considerations that are outage specific include:

- Has space been designated, footprints identified and interfacing services planned/prepared for new installations?
- Is adequate access for heavy equipment available and movement routes identified?
- Has work scope been fully defined for contracted work and are contracts executed?[*]
- Have housekeeping activities been factored into each work package?
- Have work package process documents been provided to maintenance (reliability) engineering so that they can develop equipment test plans?
- Has procurement of new equipment, major materials and outside support services—including defined delivery dates—been completed?[*]
- Collate the planned costs of the turnaround and provide estimates to management[**]

---

[*]Although these are contracting/purchasing responsibilities, it is important that planners be involved in contract review (for work scope) and in review of outage procurements in order to identify gaps between contracted work and in-house planned support and to compare work package equipment and materials lists with purchasing outage procurements.

[**]The planner's cost estimates, based on completely planned work and material costs, are an order of magnitude more accurate than estimates made by management during Phase I. These estimates must be provided to management as early as possible, in order to make a determination of whether to modify the original budget, add or delete major maintenance work or change the plant shutdown duration.

The scope of work and individual work packages for a major maintenance outage is much different from the scope of backlog work packages. For example, if older production equipment is to be replaced with new equipment, dismantling and removal of the old equipment must be completed prior to delivery of the new equipment to the installation site. The planner must carefully analyze each outage work project to identify which job activities must be completed prior to the start of follow-on activities in the work project. These work–precedence relationships between various job activities must be clearly defined. If they are not well defined, effective scheduling of labor resources will be impossible. Finding out, while in the midst of working on an outage project, that some prerequisite work must be performed before they can proceed could result in a maintenance crew sitting idle for several hours—even as long as several days!

Turnaround Checklist for Planners

❏ 1. Determine the general scope of turnaround work from engineering schedule and from previous preventive maintenance performed during shutdown.

❏ 2. Determine general labor resource requirements and contract labor.

❏ 3. Determine the general number of supervisors required.

❏ 4. Submit recommendations for required labor and supervisors.

❏ 5. Determine equipment that will be required, such as cranes, large quantities of scaffolding, compressors, welding machines or torque wrenches, etc., and if they will be available.

❏ 6. Determine status of materials, such as valves, internals, etc., and that they will arrive in sufficient time for checkout prior to use.

❏ 7. Determine pre-turnaround work for the project or other work and have work orders issued.

❏ 8. When the general scope of the turnaround is fairly stable, draw up a schedule for use at first meeting and to determine work force requirement.

❏ 9. When work force has been determined, write requests for labor.

❏ 10. Write request for equipment required that will have to be rented.

❏ 11. Write request for labor supervisors if applicable (timekeeping, etc.)

❏ 12. Submit *personnel requisition* for turnaround clerk as applicable. Note: This should be done several weeks before needed to allow for approval and recruiting.

❏ 13. Request copy machine as applicable.

❏ 14. Request additional phones; one for supervisor area, one for materials use and one in material trailer.

❏ 15. Have desks, chairs and tables moved in for coordinator, materials and zone supervisors.

❏ 16. Order portable toilets (early) if required.

❏ 17. Submit letter of request to safety for safety orientation of contract and crews and make appointments,

❏ 20. See that material and tool trailers are properly supplied.

❏ 21. See that room and transportation arrangements are made for supervisors when there is a change. Change request submitted to senior supervisors.

❏ 22. Produce schedule. Distribute at turnaround meeting. Distribute final schedule as per distribution list.

❏ 23. See that PM work orders are produced. May have to initiate work orders.

❏ 24. Assemble work orders.

❏ 25. See that an objective for the turnaround is written and that it is given to production along with copies of all work orders.

❏ 26. Produce readable copies of all work orders for trade supervisors to be included in packet to supervisors.

❏ 27. See that all work orders are activated and that all planning, including materials, is completed.

❏ 28. Arrange for transportation for crews as applicable (confer with trade supervisors).

❏ 29. Arrange with production for an area for lay-down of surplus equipment.

❏ 30. Arrange for an extra dumpster for waste.

❏ 31. Periodic update on status of preparation work and planning.

❏ 32. Provide a telephone and beeper list of personnel for the shutdown along with other frequently used numbers.

❏ 33. Secure a list of contract workers.

❏   34. Provide a typewriter, forms and office supplies for turnaround office.

❏   35. Secure forklift if required.

8.1.2.1 Purchasing: Plant Shutdown Logistics

Acquiring the parts and materials necessary to ensure shutdown success is generally a divided or fractured activity at most plants. Maintenance, production, procurement and even engineering have traditionally had a role in "chasing parts." By establishing an integrated and scheduled material management effort, accountability and systematic updates and reporting can be established in the months leading up to the shutdown. This ensures that all required material is ordered, delivery arranged according to scheduled progression and nothing is misplaced or lost. The procurement effort then becomes integral to logistics management. Procurement managers are often promoted on their ability to get things done at the lowest possible price without regard for the possible cost. For procurement specialists, parts vendors, material suppliers and contractors competing with one another is the best of all possible worlds because it lowers the price of the item or service. However, cheaper is not always better. This has been proven repeatedly by Total Productive Maintenance (TPM), Lean Manufacturing and Reliability-Centered organizations over the last 15 years. Nonetheless, the influence of lowest price continues to dictate many procurement efforts.

Key to successful shutdowns is establishing *preferred provider* relationships early in the planning process. Determine which contractors have the best record for successful execution of shutdowns, those that have proven work processes and integrated planning and scheduling procedures. Analyze which contractors have the best safety records, the lowest rework statistics and the most responsive supervisors. Once you have identified contractors that meet your criteria for partnership, invite them early into the planning process.

Do not jam up the loading dock. Sequence the delivery and distribution of material to match schedules as well as to ensure labor and storage space while checking for applicability, bagging and tagging for further distribution to staging areas. Construct a logistics distribution diagram (an overhead drawing of all routes and distribution or staging areas) and map out the flow of material. Bag or palletize peripheral material and parts (nuts, bolts, welding rod, gasket material, etc.) by work order number and supervisor. Sequence these bagged items with the major assemblies with which they will be used.

Pre-stage large assemblies at a site central to the units to be worked on. Generally, the plant warehouse is too remote from the work area to be considered a central site and has limited access. Choose a site that can be sheltered with multiple accesses and easy, but controlled, accessibility. Lay out an entrance and exit plan.

8.1.3 Phase III: Scheduling

The scheduling phase of the maintenance outage can be the success or failure determinant of the plant turnaround. The scheduler must begin his activities while the planning phase is still in progress. In plants where planning and scheduling are performed by the same person, coordination of these efforts is straightforward. Where the planners and schedulers are separate people, a continuous dialog must be established between them to ensure complete and accurate integration of their efforts. Applying traditional project management techniques for sequencing, monitoring, executing and controlling the progress of the shutdown can identify various scheduling, resources, and cost questions such as:

- Is the amount of work doable within the allotted period?
- What are the critical path jobs for completing the shutdown on schedule?
- Have enough resources (personnel, time, money) been allocated?

Beginning with the prioritized work list, the scheduler must:

- ○ determine if the work can be completed by the established cutoff date;
- ○ identify when new equipment, major materials and outside support services must be on site;
- ○ sequence work so that all in-house resources are utilized all the time and identify resource augmentation requirements;
- ○ weigh priority against job duration; it is often advisable to start longer duration jobs earlier than high-priority jobs.

Traditionally, plant shutdowns to accomplish major maintenance leave significant slack for production personnel. In the worst case, they are often faced with forced leave. In the Lean Manufacturing plant, many production line operators have been trained in performing some level of maintenance. Additionally, operations personnel, whether trained in maintenance or not, are valuable assets because they are familiar with the equipment and systems that they operate, know the plant layout, are familiar with the organizational structure and processes and, in general, possess knowledge that can be of benefit during the maintenance outage. Early in the outage-scheduling phase, schedulers should meet with operations supervisors to identify operations department resources, skills and availability. The scheduler must then integrate the available operations resources with maintenance resources to develop and staff planned work schedules. Often a skilled maintenance tradesperson can be freed-up for other work through assignment of a minimally trained line operator, who is able to perform general-purpose work following the guidance of other team members. Available maintenance staff can provide much wider job coverage when augmented by production personnel in this manner.

The scheduler must carefully consider the work location factor during schedule development. In the rush to accomplish the many tasks involved in a shutdown, schedulers often do not take into account the physical location of the work to be performed. Pipefitters are welding above millwrights, who are working above electricians working in exposed electrical panels. Such a situation provides no mechanism for the prevention of injury or damage to equipment and work is slowed while one maintenance team waits for another team to complete a conflicting activity. Schedulers who are not familiar with an area of the plant in which they are scheduling work must perform a walk-through of the area to familiarize themselves thoroughly with equipment and support system locations.

No area is more neglected during maintenance outages than clean up, before, during and after a specific job assignment. The ability of technicians to work safely and efficiently, to prevent contamination of bearings and gears, and to ensure the safe, on-time start-up of equipment are all directly related to the cleanliness of the work area before, during and after a job. Planners and schedulers alike must factor housekeeping activities into every job. If contracted work does not include maintenance of work area cleanliness (it should) within the scope of work, then in-house resources will need to be allocated to the task.

The final step in the wrap-up of any work package is test and inspection. Responsibility is assigned to one individual to inspect gears and bearings before closing to ensure no foreign objects or contamination have been left behind. Once inspected and closed, such closings should be sealed with a tamper-proof seal that will evidence unauthorized entry and the need for re-inspection. Electricians and instrument technicians make a final check of connections, proper rotation and fusing. As part of the shutdown work schedule and management plan, a test plan should be developed for all equipments prior to start up, not just those machines that were worked on. System level and interfaces, as well as individual equipment, must be tested to ensure complete restoration of services. The test plan is a responsibility of Reliability Engineering. In order to expedite development of the test plan, Maintenance Planners should provide engineering with each work package's procedural documentation. As the outage schedule is developed, the maintenance scheduler and reliability engineer will need to coordinate the scheduling and performance of the test plan requirements.

In summary, the maintenance outage considerations that schedulers will need to act on include:

- resource levels and time frame adequacy to complete the outage work package;
- logistics coordination with work schedule (material and equipment delivery/staging);
- integration of production resources with maintenance resources for work package execution;
- schedule analyzed for work location conflicts;
- essential housekeeping efforts provided for in the shutdown schedule;
- test and inspection requirements (Test Plan) factored into shutdown schedule.

8.1.4 Phase IV: Execution

The execution phase of the maintenance outage is the validation of the Planner's work packages and the Scheduler's work assignments and schedules. The first step of the execution phase, actually the final prerequisite for execution, is the Maintenance Outage Coordination Meeting. Structured much like the weekly maintenance schedule coordination meeting, this meeting is likely to last at least one full day and perhaps more. The Outage Coordination meeting should be conducted approximately one to two weeks prior to the commencement of the scheduled shutdown. All Maintenance, Production and Purchasing/Stores management and supervisory personnel should attend. Attendees should have been provided with the Outage Major Milestones Gantt Chart as well as individual project schedules applicable to each attendee at least 3 to 4 days prior to the meeting. The objectives of the Outage Coordination meeting are basically the same as those for the weekly production/maintenance coordination meetings. Any special requirements for work to be performed should be identified during the coordination meeting and not when the job starts. Similarly, any potential or actual logistics problems not yet resolved must be identified (e.g., supplier notification of parts/material nonavailability, contracts still awaiting negotiation, etc.).

Today's project management software programs are capable of utilizing several common project management methodologies such as Critical Path Method (CPM) or Project Evaluation and Review Technique (PERT). Computer technology enhancements for these classic techniques allow the shutdown scheduler a simple way to provide management with graphic presentations, resource allocation and leveling, calculating costs, communicating and delegating tasks, updating project status, reporting and analyzing "what-if" scenarios. During the execution phase, the scheduler must review work progress and work schedules several times a day. Especially important are critical path jobs (see Section 8.2 later in this chapter for an explanation of the Critical Path Method of scheduling and Appendix D for a more thorough CPM tutorial). Any perturbations to the work schedule must be identified as early as possible so that adjustments can be made prior to the need to drop one or more work items.

The role of the supervisor in the successful execution of a shutdown cannot be over-emphasized. Once all planning, scheduling, pre-staging and paperwork are complete, the supervisor must make the shutdown a reality. Supervisors, whether from the plant work force or contractor, must be trained in exactly what is expected of them. Specific roles and responsibilities applicable to the shutdown organization must be communicated. Work crews should be assigned based on the optimal span of control. Generally, each supervisor should be responsible for no more than 15 to 20 workers. This allows for hands-on interaction and follow-up on all jobs assigned to that crew.

8.1.5 Phase V: Debrief and Lessons Learned

The last phase and also a key element of shutdown success is also the first step in ensuring that your next shutdown is even more successful. Once the production equipment has been tested, equipment started up and product is rolling off the lines, it is human nature to breathe a sigh of relief and not think about shutdowns until the next time. It is also, just at this moment, when all of the successes and problems of the maintenance outage are still clear in everyone's mind, that a post-turnaround analysis and critique must be carried out.

The shutdown critique is a formal undertaking designed to root out what went right so it can be repeated, as well as what went wrong so it can be eliminated. The shutdown management plan brings together all the individual contributors to the shutdown. The critique phase reviews each player's contribution. Survey data can be collected from contractors and vendors. The managers and supervisors can interview plant work force and interpret performance data from the CMMS' shutdown project measures of performance reports. Questions that can be used to analyze the shutdown include:

- what issues helped or hindered shutdown performance?
- what was your evaluation of the shutdown organization and key staff?
- how was productivity measured? what data did you collect?

- how was performance measured? what data did you collect?
- did you meet all shutdown goals and objectives?

8.2 CRITICAL PATH METHOD SCHEDULING

One of the more important tools available to the Scheduler during a maintenance outage is the Critical Path Method (CPM) of scheduling. Whether computer automated or performed manually, CPM is the quickest and most accurate method available for scheduling and managing large, complex work packages, optimizing schedules and for modifying schedules when disruptions occur. CPM scheduling is a graphical technique used for illustrating activity sequences, together with each activity's expected duration, to portray project execution steps in precedence order.

During many large overhaul and outage situations, a project or job may consist of a number of activities that can be carried out simultaneously. The effect of doing several activities at the same time is to reduce the total time for the job to be completed. The total man-hours involved will however remain substantially the same. By tracing the various work element paths from project start to project completion, the most time-consuming path is the length of time it will take to complete the job. It is identified as the critical path because any delays along that path will delay the entire project.

Development of a CPM schedule begins by representing the project graphically by a network built up from either circles or squares and lines or arrows, which lead up to or emerge from the circles or squares. Depending on the method used, the circles or squares represent either activities or events (the completion of an activity). Connecting the circles or squares with lines or arrows represents a sequence of activities in which each one is dependent on the previous one. In other words, one activity must be completed in order to begin the next activity. The initial network development is the most important and difficult in developing CPM schedules. Graphing out the job activities and dependencies to develop the network requires intimate knowledge of the constituent parts of the project. It is remarkable how many projects are undertaken which have not been "thought through" and many persons undertaking CPM for the first time are astonished at their own ignorance of the project they are planning.

The original development of CPM scheduling technique was preceded by a very similar technique known as the Program Evaluation and Review Technique or PERT for project scheduling. Although the terms PERT and CPM today are used interchangeably, or even together as in PERT/CPM, there are in fact some subtle differences between the two methods. PERT is also known as the Activity-on-Arc (AOA) network scheduling structure. PERT became popular after it was developed and used for management of the U.S. Navy Polaris Fleet Ballistic Missile (FBM) program. It was notable in that it brought the program in within budget and 18 months ahead of schedule. Shortly thereafter, the DuPont company introduced the CPM for managing the construction and repair of its manufacturing plants. CPM utilizes the Activity-on-Node (AON) representation as opposed to the PERT AOA structure. The characteristics of the two are:

AON

- Each activity is represented by a node in the network.
- A precedence relationship between two activities is represented by an arc or link between the two.
- AON may be less error prone because it does not need "dummy" activities or arcs.

AOA

- Each activity is represented by an arc in the network.
- If activities A and B must precede activity C, there are two arcs, A and B, leading into arc C. Thus, the nodes, the points where arcs A and B join arc C, represent events or "milestones" rather than activities (e.g., "finished activities A and B"). Dummy activities of zero duration may be required to represent precedence relationships accurately.
- AOA historically has been more popular, perhaps because of its similarity to Gantt Chart Schedules used by most project management software.

A brief example will illustrate the subtle differences in the two methods. Suppose the project to be managed is the installation of a new production processing system consisting of a large, 15-ton automated metal forming and joining unit outputting to a 40-ft long steel roller conveyance, where the unit is cleaned and prepared for galvanizing, that feeds a heated (to 850°F) galvanic coating vat. The project steps for installation are determined to be (greatly simplified for the sake of this example) as shown in Table 8-1.

When depicting AOA or PERT graphics, the custom is to identify the activity above the activity line and the duration below the activity line. Therefore, the project defined in Table 8-1 when represented by an Activity-on-Arc network yields the diagram shown in Figure 8-2.

Note the dotted line leading into the activity line "LnG." Table 8-1 indicates that activities *Serv* and *Ins* are immediate predecessors of activity *LnG* but *Serv* time has already been accounted for between activities *Fp* and *LnU*. Therefore, in order to depict the paths accurately, the dummy activity labeled

Table 8-1
Steps in Sample Project

Activity	Code	Time	Predecessors
Locate and clear footprint for the 3 units	Fp	2	-
Stub out hydraulic and electrical services to units	Serv	5	Fp
Reinforce floor under fabrication unit	ReU	3	Fp
Reinforce floor under galvanic vat	ReG	3	Fp
Install insulated standoff beneath galvanic vat	Ins	1	ReG
Land and connect services to fabrication unit	LnU	2	Serv, ReU
Land, fill and connect services to galvanic vat	LnG	3	Serv, Ins
Land conveyor section	LnC	1	LnU, LnG
Connect and align conveyor to both units	C & A	2	LnC

None with duration of 0 must be added to the figure. In this case, it is critical to show the dummy activity because it is in the critical path for completion of this project. The critical path in Figure 8-2 is represented by double lines.

When depicting activities on an Activity-On-Node or CPM networks, the activities and their duration are shown inside the node. The same project

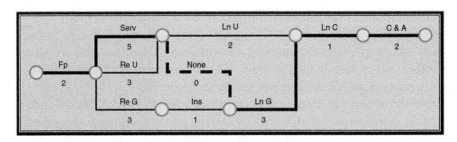

Figure 8-2 Activity-on-Arc (AOA) Network Diagram

from Table 8-1, when depicted as an AON network yields the diagram shown in Figure 8-3. There are a number of different conventions for depicting the critical path on CPM networks. Figure 8-3 illustrates several of them, however only one need be used. Node connectors with arrowheads, nodes with double lines, different color nodes and different color lettering are all shown here.

Looking again at Figure 8-3, tracing the path Fp → Serv → LnG, you see that LnC cannot occur before day 10. Alternatively, if you trace the path to LnC as Fp → ReG → Ins → LnG, you see that it requires only 9 days. Looking at this second, shorter path, you can see that any activity not in the first, or critical, path can be delayed up to one day without impacting the start of LnC. This is called "float" or job slack. Float defines when and for how long it may be possible to divert resources to another job. Such diversion should be approached with extreme caution because the likelihood of two jobs precisely following their planned durations and/or meeting their planned start dates is very low. However, float does define points where the scheduler might look during the outage for labor resources to meet emergent, short-term needs.

To facilitate the use of CPM, creation of a shorthand notation scheme is advisable so that longer, descriptive phrases would not need to be written into the symbols used in CPM scheduling. The Work Breakdown Structure (WBS) is the most common tool used to create this shorthand notation. The WBS is a hierarchic breakdown of a planned work package into successive levels, each level being a further breakdown of the preceding one. Each item at a specific level of a WBS is numbered consecutively (e.g., 10, 10, 30, 40, 50) and each item at the next level is subnumbered using the number of its parent item (e.g., 10.1, 10.2, 10.3, 10.4). The WBS may be drawn in a diagrammatic form (if automated tools are available) or in a tree-form resembling a Microsoft Windows™ folder-file tree.

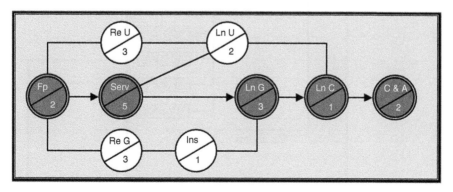

Figure 8-3 Activity-on-Node (AON) Network Diagram

The WBS begins with a single overall task (level 1) representing the totality of one planned work package. Figure 8-4 illustrates a maintenance job (10) with 2 level 2 tasks (10.1 and 10.2), one of which (10.1) has 2 level 3 tasks (10.1.1 and 10.1.2), one that has no subtasks and a second one (10.1.2) that has four sub-tasks (10.1.2.1 through 10.1.2.4).

In larger plants, the small investment cost for CPM scheduling software will be returned many times over in saved time for schedulers. Many CPM-based project management programs provide a variety of options for depicting WBS, CPM Network layout and even the node format and node data contents. In Figure 8-5, for example, the CPM Node can contain as many as seven data fields. These programs will automatically depict critical paths and some will generate an entire shutdown schedule once the individual projects or jobs have been set up.

Once the critical path length for a project has been identified, the next question invariably asked is, "can we shorten the project?" The process of decreasing the duration of a project or activity is commonly called crashing. For many construction projects, it is common for the customer to pay an incentive to the contractor for finishing the project in a shorter length of time. For major maintenance work, it may be possible to shorten the duration of an activity by allotting more resources to it, adding an extra shift or

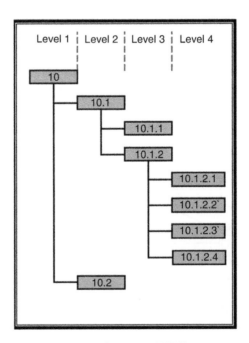

Figure 8-4 Simple Work Breakdown Structure (WBS)

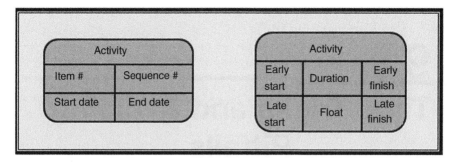

Figure 8-5 CPM Node Data Contents

even procuring a different unit to accomplish the job's objectives. Whenever consideration is given to "crashing" a critical path task, a careful cost analysis of the alternatives involved must be performed. The cost savings of shortening the plant shutdown by one day may, or may not, offset the cost of an additional shift for 3 or 4 days. Make informed decisions. Refer to Appendix D for a brief, but more detailed, tutorial on construction, analysis and utilization of graphical scheduling techniques with Gantt Charts and CPM Diagrams.

9

Tips, Tricks and Avoiding Pitfalls

9.1 STRAIGHT FROM THE HORSE'S . . .

Even though Maintenance Planning and Scheduling has not been around that long, it does not take long to make mistakes, and without mistakes, one cannot learn from them. This chapter has gathered together a smattering of lessons learned, often the hard way, from practioners of the fine arts of Maintenance Planning and Scheduling. Although it is human nature to at least try to avoid making errors, especially in one's chosen vocation, they are bound to make their way into your operation, often at the most inopportune time possible. Do not be discouraged, rather look on it as a necessary part of attaining perfection and above all, *learn* from what doesn't work. Oh yes, and don't forget to learn from what does work as well, because this chapter also contains many tips and tricks of the trade garnered from, sometimes unexpected, successes.

9.2 GAINING TRUST

9.2.1 Maintenance Planner

For a variety of reasons, some of which will be discussed in the paragraphs that follow, maintenance trades tend to mistrust maintenance planners. Unless planners can overcome this and win the trust of the maintenance trades, theirs will be a long and arduous journey toward having their work plans carried out effectively. There are, in fact, some proven methods to overcome many of the inherent reasons for mistrust and the planner, who applies common sense towards understanding the human condition will be able to win the hearts and minds of most, if not all, of the maintenance trades with whom he or she is involved.

There is less sense of achievement when plans for carrying out day to-day work activities are prepared and assigned by others. When planning for others, it becomes most important to clarify objectives. Plans need to be designed which provide satisfactory signals when those objectives are being reached. A means for providing feedback of success or failure must be built into job plans prepared for others. Performance of individual members working as a group improves the most when they receive constructive information about their individual efforts as well as the group's success as a whole, particularly if the jobs and work are difficult and/or complex. The job of maintenance planning, even more so than scheduling, must be approached as if its success depended on the success of those executing the plans because in fact it does. One method to build rapport with maintenance groups is for the planner to request, via the supervisor, assistance from one of the group to plan for a particularly broad-in-scope work order. This may not always be possible due to the overall workload, but if and when the opportunity arises, take advantage of it. This kind of work sharing goes a long way towards recognition of the maintenance planner as part of the team.

Nearly everyone is convinced that their own methods, or plans, are better than someone else's methods and plans. This conviction is to some extent acted out; that is, greater effort is expended to ensure that one's own judgments are validated by the outcomes. It is as if people were saying, "I think this will happen, therefore, I will make it happen." The confidence of planners in the adequacy of their plans can be communicated, while any reservations on the part of the trades need to be brought into the open and discussed with the planners. If the doers believe that the plans are unrealistic, even if the plans are actually sound, the doers may behave in a way that confirms their own beliefs. Planners need to share with the doers the reasons for their optimistic expectations.

Persons not actually involved in the development of plans are less committed to making those plans work. One method for overcoming this is to consult with maintenance trades at various stages in the planning process. Wherever possible and suitable, the ideas of the trades should be incorporated into the plans, conversely when such ideas are unusable, the reasons should be discussed with trades. Job plans may also allow for some discretion action on the part of the technician to modify noncritical elements of the plans, thus increasing the feeling that the maintenance trades have some control over the fate of the plans, and consequently, a level of responsibility for their successful execution. Provisions should be built into the planning process to obtain periodic feedback and planning evaluation from the trades.

Maintenance Groups report greater difficulty in understanding plans that others have assigned to them than they find with their own plans. Understanding of the plan can be fostered by ensuring that the plan itself has been created in a way to minimize its ambiguities. Repeating instructions may increase reliability and understanding. Crosschecks and tests of the plan's clarity before it is presented to the maintenance crews may be helpful. Additionally, the plan's instructions need to be simple enough to be understood by the least capable within the maintenance crew, who usually ends up reading and interpreting the plan's instructions for the others. If he can understand the stages of the plan, all of the others can also understand it.

9.2.2 Maintenance Scheduler

Due to the fact that schedulers spend more of their time coordinating equipment availability needs with operations and most of their direct contact with maintenance groups is via the group(s) supervisor, the trust factor does not impact the scheduler's job performance nearly as much as it does the planner's performance. However, there is some degree of mistrust for anyone outside of the group who is, in some form or fashion, providing direction of the group's activities.

Approaches similar to those recommended for the maintenance planner will quickly overcome doubts that the trades might have relative to the scheduler's knowledge and abilities. Frequent visits by the scheduler to work sites for validating job duration estimates and obtaining instantaneous feedback will help enormously. The scheduler's objective, just as the planner's, is to become looked at as an integral part of the team.

The inherent mistrust of the "outsider" no longer exists when the scheduler is recognized as a team member who is interested in helping maintenance workers perform more effectively. It is not likely that maintenance trades will volunteer information, unless they have a real "beef" with the scheduler, so the scheduler must actively draw out that information. Ask the maintenance trades about the sufficiency of the time allotted for scheduled work. Find out if operations has met their commitment to make the equipment available as scheduled. Were there any problems encountered moving from one job to the next scheduled job? When querying the maintenance staff, make reference to "we" in maintenance and "they" in operations– "did *we* have access to the equipment when *they* promised it?"–to reinforce the maintenance team member image; at the same time exercise moderation to avoid depicting operations as the enemy.

9.2.3 Attitudes, Practices and Methodologies for Success

9.2.3.1 Empowerment and Areas of Responsibility

Empowering maintenance groups does not mean that they are completely left to their own devices, free to pursue their own objectives; rather instead, empowerment means, "they are free to make decisions for those matters that are within their areas of responsibility." Mastery of the maintenance process requires a well-planned and coordinated *team* effort; therefore explicit job planning and work scheduling processes are necessary to improve trade productivity. Superior, effective maintenance requires maintenance trade empowerment *and* it requires maintenance planning and scheduling.

Lean Maintenance, or any Lean Operation, is about eliminating waste. It is about identifying and eliminating those activities that do not add value to a process. In manufacturing plants that do not utilize a maintenance planner/scheduler, the average tradesperson only spends about 3 hours, or less, working on equipment (wrench time) during an 8-hour shift. The other 5 hours, or more, are spent on activities such as obtaining parts, traveling around the plant site, receiving job instructions or even waiting to be assigned another job. In those environments, maintenance trades must accomplish these nonproductive activities in order to complete maintenance work even though they add no value to the maintenance process.

One part of the problem is that maintenance group supervisors have traditionally been granted a significant degree of freedom in selecting work activities and defining maintenance productivity. Additionally, while maintenance managers have been strongly supportive of the concept of empowerment, they have often allowed the term to define an extreme level of autonomy. As a result, when maintenance planning and scheduling is instituted, the maintenance supervisor clearly sees his autonomy, and authority, being substantially diminished. This attitude can create an insurmountable barrier to acceptance of the persona of the planning and scheduling functions by the maintenance trades. The solution can be recognized by returning to the definition of empowerment as "free to make decisions for those matters that are within their areas of responsibility." Obviously, the areas of responsibility need to be well defined, where previously, they were more than likely undefined.

It is necessary to begin by defining some of the necessary components of the maintenance process, which in turn will help define the proper area of responsibility of a maintenance group and supervisor and thus deal with the concern of empowerment. First, the maintenance process makes considerable use of the WO system. By pointing out that, in actuality, the work order is the vehicle by which requesters of work identify needs; maintenance planners predict parts, tools, skill levels and hours; and the maintenance group executes work. Clearly, the group supervisor does not "lose empowerment" by the work order being processed among various groups. It is the established priority and criticality system that communicates the importance of individual jobs. In fact, the system really drives, to a large degree, which jobs are next scheduled for the maintenance group. Again, it should be pointed out to the group supervisor that he or she should view this system as helping, not hindering, his or her job performance. The similarity that the group supervisor does not feel it necessary to have complete control over hiring, training, tools, spare parts and payroll can also be pointed out to help alleviate any feeling of diminished control.

In this situation, the fundamental dividing line that should be used to define areas of responsibility is time. The planning and scheduling group works in the *future*, while the maintenance group and the group supervisor work in the *present*. Planners develop job plans with cost and time estimates and they identify the parts, materials, tools and other requirements that *will* be needed when the work is performed. Schedulers develop the weekly schedule by looking at the entirety of available backlog and they use resource reports of labor availability without regard to individual names to determine how much work *will* be performed. The planning and scheduling group executes its portion of the maintenance process for the overall benefit of the maintenance team (the entire Maintenance Department) under the direction of the maintenance manager. This is the proper area of responsibility for the planners and schedulers.

The maintenance group trades execute assigned work, empowered to concentrate on *today's* work without regard to organizing details for future work. The group supervisor assigns and monitors *today's* work. The supervisor develops *today's* schedule using the weekly schedule and production windows of equipment availability while, at the same time, considering any emergent, high-priority work that cannot be put off to the future. The supervisor works with individual technicians. The trades and their supervisor execute their portion of the maintenance process for the overall benefit of the maintenance team (the entire Maintenance Department) under the direction of the maintenance manager. This is the proper area of responsibility for the technicians and their supervisor.

Theory suggests that the role of the technician in the planning process should be minimized. Reality, on the other hand, shows repeatedly that the maintenance technician will become involved and provides real benefit through his or her involvement with and knowledge of the planning process.

Some jobs, either by design or through default, are not covered by the planner. On these jobs, it is significantly preferable for the technician to perform the work following his or her own thought-out planning process.

Rather than charging into a job without benefit of any planning whatsoever, a knowledgeable technician will think through the entire job before he or she proceeds. This kind of forethought is particularly important on short duration and short lead-time jobs. Accordingly, it is extremely beneficial to train maintenance trades in planning basics. With the basics ingrained, even emergency work orders will gain some benefit of planning.

There are also those situations when the unavailable backlog—work still pending planning—becomes out of control. In such cases, the use of selected trades to aid in regaining control is highly recommended. It is certainly much better to use a small portion of the maintenance staff as temporary planners than it is to completely lose control of the backlog and force the entire maintenance workforce to work without benefit of prior planning, which can only lead to an ever-increasing spiral of uncontrolled backlog.

Lastly, there are situations where the knowledge of a given tradesperson simply exceeds that of either the planners or supervisors. It would indeed be foolish not to draw upon the best available knowledge when production downtime is the alternative.

Maintenance trades can be utilized as planning assistants on a temporary, rotating basis. This approach is particularly appropriate when planner support spans are excessive due to budgetary limitations. A number of organizations have used this approach with such outstanding results that they have made it a standard practice. A normal rotation period is from three to six months.

9.2.3.1 Skill Levels and Training

Although it is not unusual for maintenance personnel to have operating experience, it is not as common for operating personnel to have maintenance experience. Many maintenance technicians began their careers as production line operators and worked their way into the maintenance department. On the other hand, few, if any, operating personnel began their careers in the maintenance department and worked their way into the operations department. This issue becomes important when scheduling maintenance tasks involving both operating and maintenance personnel.

When operating personnel assist in performing maintenance tasks, what duties are they assigned? Do they merely assist maintenance personnel or are they responsible for performing key elements of the task? Are they expected to use equipment with which they are unfamiliar? In the Lean Maintenance Environment, essential training in the performance of routine, minor maintenance work is provided for the autonomous operators. Planners and Schedulers often mistakenly assume that since the person has many years of experience and has received maintenance training, he or she knows how to use the tools and equipment to perform the various maintenance tasks assigned to them. Is that assumption a safe one? Has an analysis of their experience and skill level for performing various levels of maintenance work been performed?

Analyzing the experience and skill levels of those, both maintenance and operations staff, who will be performing the maintenance work, is a key step in the planning/scheduling process and a step that is often omitted. If three workers are required to perform a task, two may have performed it numerous times and require only a minimal review before starting. However, the third person may not have the same knowledge and skill level and could require extensive additional briefing/training before being assigned to the task. The skill category and level of every labor resource available, maintenance and operations, to the maintenance scheduler, as well as the maintenance group supervisor, must be known for maintenance planning and scheduling to be successful and efficient.

Reliability Engineering should perform a skills assessment of all maintenance trades as well as those operations operators who will be available assets for executing planned work packages. It is important that the reliability engineering staff work closely with planners and schedulers to define skill classification level and categories. The planner, when developing the work package should indicate both the category (mechanical, electrical, etc.) and skill level of labor resources required to execute the work package. The scheduler, when working with the maintenance group supervisor to assign specific resources, must identify recurring shortfalls as well as excesses by category and level, to maintenance engineering so that any additionally required training can be provided.

In today's rapidly expanding production technology, training for multi-skilled resources is almost a mandatory requirement. Multiskilling is the process of training maintenance trades in specific skills that cross traditional trade lines. The advantage of multiskilling is that particular jobs that have historically required more than one trade, but not necessarily more than one individual, can now be performed by just one person, and scheduling with multi-skilled resources provides a flexibility factor that can render the scheduler's job at least ten-fold less complex.

A typical example is the change out of a small motor. Traditionally, a change-out could require an electrician to disconnect the motor leads and a millwright or mechanic to disconnect the coupling, physically replace the motor and perform the alignment. The electrician would then return to the job, reconnect the motor leads, check and possibly change rotation. The mechanic or millwright would, at this point, be able to connect the coupling halves to complete the job.

In fact, no more than one individual should be required on this job at any time, but trade distinctions often require the close scheduling of multiple trades. If the loss of this motor created downtime, both individuals would remain at the job site, performing only their particular job functions as needed. In multiskilling, individuals would receive additional training, beyond the normal skills required for their trade. The mechanic or millwright would be trained in the proper disconnecting and reconnecting of the motor leads, as well as how to change motor rotation. The electrician, in turn, would be trained in coupling disassembly and reassembly, as well as alignment methods. After this training, either individual would be qualified to perform the entire job alone. The advantage to the company in multiskilling comes with the ease of scheduling work that, in the past, required two or more trades or skill distinctions. The advantage to the worker is usually an incremental increase in pay for the additional skills learned and used.

9.2.4 Tips and Tricks of Successful Planners and Schedulers

9.2.4.1 Maintenance Planner

Contained here are practices that can enhance the work planning function and additional general considerations to keep in mind when setting up your plant's planning function as well as when executing the planning function.

1. Develop a standard form/format to use for work instructions; it can also serve as a checklist to ensure applicable elements have all been considered.
2. Utilize a small digital camera on work site walk-through so that any important or unusual conditions can be documented and added to the work instruction.
3. When identifying skill levels required to perform the work, identify the lowest level capable. This requires a solid communications link with Maintenance Group Supervisors.
4. On jobs of more than two hours duration, break the work into sequential activities wherever possible, so that specific trade/skill level requirements can be designated by activity, thus freeing up a resource for other, near-by work when not required to be on the first job. Note that when P & S are performed by different people, a method of coding or highlighting the "activity only" requirement must be developed so that the scheduler can provide for "near-by work."

5. It is a common problem for planners to become bogged down in the details of job planning to the point that many of their responsibilities are neglected due to attempts at perfecting the work instruction. The maintenance planner should define some sort of schedule for allocating his time to all of the activities necessary to perform the job. The following is representative of how much time should be devoted to the various activities and is typical of a planner's daily schedule:

Planner's Daily Activities Schedule

	Planner Activity	% of Day
1.	Job Screening	5
2.	Job Requirements	10
3.	Job Research	5
4.	Detailed Job Planning	20
5.	Job Preparation	5
6.	Procurement	20
7.	Job Scheduling	15
8.	Daily Schedule Adjustments	5
9.	Job Close Out	5
10.	Personal and Miscellaneous	10
	Total	100

Bear in mind that each and every day will not break down to these proportions, but over the course of each work week, the planner's actual daily time/activity numbers should be close to the schedule's time/activity values shown above. Analysis of imbalances could indicate a need for additional training, improved Planner Libraries, additional planners or expanded/improved use of computer databases.

6. In long established plants, be alert for possible "aging work force" problems. If system/equipment "experts" are nearing retirement, the planner must do everything possible to capture their knowledge. Well-seasoned maintenance veterans are intimate with their equipment and can quickly repair equipment to avoid downtime. Their knowledge includes key equipment condition indicators, inspection techniques and criteria and general know-how pertaining to the maintenance of the plant's production assets. This critical knowledge is often just stored in the employee's memory or sometimes written down in "quick reference" notebooks. The ideal repository to capture this knowledge are the CMMS/EAM databases and/or Reliability software, but the planner must actively pursue capturing this knowledge, perhaps even to the extent of convincing the Maintenance Manager to assign the "experts" to the planner for some period each week.

9.2.5 Maintenance Scheduler

Contained here are methods for performing scheduling and general considerations to keep in mind when setting up your scheduling function as well as when executing the scheduling function.

1. Invest in Project Management software. The cost of US$200 to US$500 will be returned in time-saved on the scheduling of just three or four complex jobs. It also provides a repository for skill profiles and similar resource data. Additionally, it will allow the Maintenance Administrative Clerk to perform some of the preliminary scheduling work.
2. Carefully select the most logical schedule week (e.g., Friday through Thursday).
3. Communicate, collaborate and coordinate with the customer. Relate the maintenance schedule to the operating schedule. They should be mutually linked and supportive.
4. Plan strategy on weekly basis; finalize tactics on daily basis.
   a. Operations and maintenance to communicate priorities and changes in prioritiesconstantly.
   b. List jobs in descending order of importance until all available labor-hours are committed.
      i. PMs listed first
      ii. Determine most logical time of day to schedule Planned Jobs, e.g., longer jobs scheduled early in the week and early in the day.
   c. Keep majority of trades scheduled to important work that should be started and completed without interruption.
   d. Assign jobs that can be interrupted, delayed, etc. to "a few good men" who are flexible. This means they can stop and resume jobs, be reinstructed and reassigned to "mini-emergencies" with minimal loss of efficiency and without morale drop.
5. Align personnel with jobs on the basis of knowledge and aptitude, while also considering individual training needs.
   a. Experience shows who is skillful in certain job types and who needs more exposure to them.
   b. Balance equipment specialization with broad facility knowledge.
   c. Utilize individual skills to the greatest extent possible. Provide tradespersons with a challenging environment and the opportunity to grow.
6. Schedule what *can* be done, not necessarily what needs to be done.
7. Distinguish between duration and labor-hours. Indicate both to make the schedule perfectly clear. It can be extremely beneficial as well as a significant time saver to make use of a versatile project management software system. Software capable of generating both Gantt Charts

and CPM Diagrams can usually be obtained for US$1000 or less. The investment can pay for itself within the first few months of use.

8. Note on schedule when shutdowns and support functions are scheduled/required.

9. Schedule multiperson/multitrade jobs as the first job in the morning whenever possible. This assures everyone is available to start the job at the same time. When scheduling multiperson/multitrade jobs later in the day, consider their previous assignment.

10. Do not assign one tradesperson to a one-hour job and the helper to a two-hour job. No way will they both be available to start the two-person job concurrently.

11. Think about crew balancing delays and plant location on multiperson jobs. Often another small job in the same area can be worked concurrently by the same crew using multiperson balancing delays. Seldom are all three members of a crew required continuously throughout the job duration. (Refer to Tip No. 4 under Maintenance Planner.)

12. Avoid duplicate shutdowns by assembling all work requiring common equipment access.

13. Save minor indoor jobs for extreme temperatures and inclement weather.

14. Eliminate unnecessary trips by sequencing work by location.

15. Follow schedule progress and make adjustments as dictated by "real" operating needs and changing resource constraints. Valid considerations are:

    a. Schedule adherence/protection *is* an effective use of overtime.

    b. Give daily schedule priority to work orders scheduled for the previous day but not completed.

    c. Sometimes the schedule should be sacrificed to take advantage of downtime windows of opportunity for higher priority work orders.

16. Always be sure the job site or machine is available to the scheduled crew by checking with operations just prior to scheduled availability. Additionally, check with the scheduled crew to verify timely arrival at the work site.

17. Ensure that schedule is studied and approved by everyone concerned. Approval means that a contract has been reached between operations and maintenance to comply with "their joint schedule" for the deployment of maintenance resources in support of operations plans. Point out any unusual requirements that might be overlooked—e.g., equipment requires shutdown one hour prior to work commencing for cool down.

10

Metrics: Measuring Planning and Scheduling Performance

We often refer to metrics, which is just a term meaning "to measure" (either a process or a result). A Performance Measure, or metric, is simply the measurement of a parameter of interest, such as "Labor Hours Scheduled." The combinations of several metrics yield indicators, which serve to highlight some condition or highlight a question that we need an answer to, such as "Completed Hours ÷ Scheduled Hours." Key Performance Indicators (KPIs) combine several metrics and indicators to yield an assessment of critical or key processes.

Any endeavor needs to be able to define not only where it is headed, in terms of stated goals and objectives, but also to determine the progress being made towards those goals. This requires defining the KPI's that need to be monitored in order to gauge progress towards those goals. One of the more interesting points here is that KPIs can be created in a hierarchical and interlinked fashion which allows management to pinpoint the root causes of system failures. In order to determine maintenance planning and scheduling strengths and weaknesses, KPIs should be broken down into those areas for which you need to know the performance levels. Depending on KPI values we classify them as either leading or lagging indicators.

Leading indicators are indicators that measure performance before a problem arises. To illustrate this, think of a key performance indicator as you drive your car down a road. As you drive, you deviate from the driving lane and veer onto the shoulder of the road. The tires running over the "out of lane" indicators (typically a rough or "corrugated" section of pavement at the side of the road that serves to alert you to return to the driving lane before you veer completely off the pavement onto the shoulder of the road). These "out of lane" indicators are the KPI that you are approaching a critical condition or problem. Your action is to correct your steering to bring you car back into the driving lane before you go off the road (proactive condition).

If you did not have the indicators on the pavement edge, you would not be alerted to the impending crisis and you could veer so far out of the driving lane that you end up in the ditch. The condition of your car, sharply listing on the slope of the ditch, is a lagging indicator. Now you must call a wrecker to get you out of the ditch (reactive condition). Lagging indicators, such as your budget, yield reliability issues, which will result in capacity issues.

10.1 PLANNING AND SCHEDULING PERFORMANCE INDICATORS

10.1.1 Labor Performance

Accuracy of job plans and achievability of work schedules have a tremendous effect on maintenance labor performance. Of course, the skills, work ethics, supervision quality, maintenance stores and a number of other factors also impact maintenance trades performance levels. Labor performance measures can be trended over time to provide an indication of improvement levels of all contributing factors, in particular a recent start-up of the planning/scheduling functions. Labor performance can be in the form of:

- Labor Efficiency is the traditional efficiency report calculation (standard or estimated labor hours divided by actual labor hours). The measure indicates how well the crew is performing in relation to the expectation. Variance from estimate might indicate poor crew performance, inaccurate estimate, or inaccurate work scope. The labor efficiency number as a percentage is:
- (Total Estimated Hours for Work Orders Completed/Total Actual Labor Hours for the Same Work Orders) × 100
- *Schedule Compliance* = Scheduled Work Completed (hours) ÷ Scheduled Work (hours).

Schedule Compliance is merely a number that indicates a quantitative value of work completed vs. work scheduled. While important, a more qualitative report that considers the "whys" of nonconformance is more important to more people. In order to provide the most accurate compliance number, the hours of scheduled work used in the calculation should be the sum of the adjusted daily scheduled hours for each workweek.

Calculations should be made weekly and should be made at the supervisory level—all crews responsible to given supervisor or foreman. The one-week period puts several jobs into the calculation, allowing for an averaging of unusually difficult job occurrences with unusually easy occurrences. Accuracy on individual jobs is not reliable, but accuracy over the several jobs completed by a crew on a given week is acceptably accurate: particularly for measuring performance trends.

Individual (as opposed to crew) performance can be calculated periodically, but only to guide necessary training to the critical needs and never for disciplinary reasons. The use of work measurement for discipline detracts from more important applications such as backlog control, work scheduling and group performance trends.

10.1.2 Job Planning and Scheduling

The planning and scheduling functions are inextricably linked to each other in their effect on overall maintenance effectiveness. As a result, it is often difficult to gauge the planning effort effectiveness and the scheduling effort effectiveness separately. There are a few measures and indicators that are predominantly influenced by one or the other. In order to gauge planning accuracy, measurement of the cost of completed, planned work together with completed work order data can provide:

- *Planning Accuracy* = (# WOs completed within estimate [±15%]) ÷ (# WOs completed)

On the other hand, without regard for estimates of cost and time, the amount of pure "wrench time" is mostly determined by work scheduling effectiveness. Combining payroll data with completed work order data can provide:

- *Labor Utilization* = Labor Productive Time (wrench time) ÷ Labor Paid Time   The target percentage is 80%

The remaining measures can be dominated by either planning effectiveness or scheduling effectiveness or can be influenced equally by them as well as by factors outside the planning/scheduling functions. Nonetheless, because the P/S influence exists, the following indicators should also be monitored and trended:

- Schedule Effectiveness and Schedule Performance are two additional scheduling metrics that assist Maintenance Management to fully understand how their labor resources are being utilized. The data for these metrics comes from the same sources as Schedule Compliance and Labor Efficiency.

- The schedule effectiveness number as a percentage is:
  (Direct Scheduled Hours Completed/Labor Hours Available) × 100
  The target percentage is 65%
- The schedule performance number as a percentage is:
  (Scheduled Hours Complete (Direct & Indirect)/Labor Hours Available) × 100   The target percentage is 80%
- Another powerful metric of tracking performance of the scheduling process captures the percent of work that was planned for the scheduled week. Trending this metric on a weekly basis provides a clear picture of the relationship of labor planned versus breakdowns.

The percent planned work number as a percentage is:
(Hours on Planned Work/Total Hour Worked) × 100

Additionally, there are several general maintenance operation indicators that are influenced by the P/S function to a greater or lesser extent depending on the maturity of the function. These are generally monitored by maintenance management as overall indicators of maintenance department effectiveness, but may be used to provide additional refinement of the P/S performance metric.

- OEE = $A \times P \times Q$

where OEE is Overall Equipment Effectiveness, and A is Asset Availability, P is Asset Performance and Q is Quality of Output.

- *Maintenance Cost Per Unit Produced* = Number Units Produced ÷ Total Maintenance Costs (for a given report period)

10.2 RELIABILITY EXCELLENCE (Rx) AND THE P/S FUNCTION

Rx is that state of maintenance management and performance that *effectively* applies the leading edge policies, procedures, systems, structures, methods and technologies available for the maintenance function. The keys to achieving Rx do not involve any new concepts. They have always been and still remain, "get the basics right and make continuous improvement towards Rx a goal of the entire organization." At the very top end of the Rx level of operations is the *World Class* operation. World Class refers to that state of readiness and application that classifies an organizational entity (business, operation, function) competitive with "the best of the best" in the world.

There are many common attributes possessed by World Class maintenance operations. The following are some of the most prevalent and are also considered to be some of the most important:

1. management awareness of, commitment to and appreciation for the importance of maintenance in order to realize the total organization's overall objectives and vision;
2. focus on service to the internal customer;
3. a proactive, rather than reactive, approach;
4. equipment stewardship and custodianship;
5. focus on root causes analysis, knowledge of machine condition and first time quality of the ultimate product;
6. maintenance/production partnership;
7. participative, Empowered Teams;
8. planned blurring of traditional interfunctional barriers and job jurisdictions;
9. operator participation in preventive maintenance inspections and other routine maintenance activities;
10. self-motivation and direction (empowerment);
11. continual training, cross training and multiple skills development;
12. continuous process improvement through a value-added focus;
13. information sharing;
14. benchmarking;
15. effective application of advancing technology.

The "quest" must be for Reliability Excellence at the World Class level, rather than merely reduction of maintenance costs.

As organizations pursue a World Class level of operation, they need comparative measures showing their own progress relative to progressive, highly regarded organizations, which have earned their reputation for being world class. Such organizations focus on customer service (internal as well as external), first-time quality and highly productive work teams. For the Maintenance Planning and Scheduling functions, the comparative measures are those delineated in Section 10.1.

There are no magic solutions, no short cuts to developing the ideal maintenance program. In order to obtain the highest return on investment, a comprehensive program encompassing all variables relating to maintenance must be developed. The optimum maintenance operation is a result of serious, continuing attention given to a large number of items, many of which involve not only maintenance but also: management, the internal customer and those organizations to which maintenance is the internal customer.

Often, before a program can be successfully launched, management must first be acquainted, even educated, about the relationship between quality and maintenance.

10.3 SUMMARY: CHARACTERISTICS OF RELIABILITY EXCELLENCE

In order for a plant to operate at the level of Rx, the maintenance organization and the plant's supporting structure for the maintenance operation, should display these characteristics:

1. A set of "Shared Beliefs" is apparent throughout the organization, thereby enabling the *"Cultural Transition"* from Reactive to Proactive Maintenance, in support of *"World-Class Operations."*
   a. Organizational focus is on Rx. Maintenance cost reduction will ultimately become a by-product, but in the short term must not undermine the long-term viability of the maintenance operation. Despite short-term urgencies, World-Class organizations do not lose sight of the long-term mission, objectives and goals.
   b. A "Maintenance Mission Statement" is prominently displayed and emphasized to assure a well-ingrained vision with sustained understanding, commitment and support of the proactive Rx process. The maintenance function is responsible for *Maintenance & Repair (M&R)*. Although Rapid Response/Repair of true emergency and urgent conditions (unplanned and unscheduled line support) is necessary in support of daily production, it is Important Proactive Maintenance (planned and scheduled backlog relief and Preventive Maintenance) that ultimately improves Asset Reliability and thereby preserves long-term viability of the maintenance operation.
   c. A well-conceived "Master Plan for Maintenance Improvement" exists with actual progress charted relative to the plan. The program is holistic, integrated, cooperative, participative and proactive with continual emphasis upon *The Basics*.
   d. The Maintenance Organization exhibits Functional Professionalism and Marketing. Thereby, Maintenance is recognized as a "Contributing Resource Center" and "Proactive Messages" are continually conveyed throughout the organization.
2. There is every indication that the chosen *"CMMS/EAM"* is fully functional, including all modules. Workload cannot be defined without benefit of an Effective Work Order System.

3. Equipment Configuration is complete. Equipment and Material Databases are fully loaded and reference documentation (prints, procedures, etc.) is available on-line.

4. Evidence exists that *"Cost Distribution"* to Work Orders is complete and accurate (Labor, Material, Contract Services).

5. *"Work Programs"* define the balance between resources and workload. Hourly as well as salaried resources must be kept in balance with workload to avoid deferred maintenance and ineffective functional management, each of which perpetuates the state of reactive/unreliable maintenance.

6. A well-conceived and quantified *"Organizational Structure"* is established by which proper distribution of resources to the three forms of maintenance is achieved.

7. Clear *"Job Descriptions"* and Responsibilities (hourly and salaried) are established and available for the benefit of all parties.

8. The *"Baseline of Critical Maintenance Measures"* is defined and quantified with relative improvement routinely measured (reference the iceberg). Elements of cost and profit that can be favorably impacted by improved maintenance are identified and measured.

9. Quantified *"Goals"* with progressive interim *"Targets"* (milestones) relative to the Baseline Measurements are established and quantified.

10. Work is performed efficiently as a result of organized planning, effective material support plus coordinated, scheduled and well led work execution.

11. Well-defined *"Job Plans"* are established, cataloged, issued and utilized/referenced.

12. *"Weekly Schedules by crew, day, individual and job"* are developed, adhered to and monitored in cooperative, coordinated partnership with the internal customer. Thereby, effective application and utilization of all maintenance resources as well as "Service to the Internal Customer" are assured.

13. *"Equipment Uptime"* is Continually Improving through the establishment of a reliable Preventive Maintenance program designed, directed and monitored by strong Maintenance Engineering support. *"Maintenance Engineering"* continually refines the *"PM Routines"* through application of *"RCM, Root Cause Analysis, Maintenance Optimization and Machine Condition Based Maintenance"* wherever feasible and the PM Program is conducted as a "Controlled Experiment" with *"Programmed Equipment Access"* providing for *"Proactive Maintenance."* Continued refinement is documented and quantified.

14. It is evident that *"Equipment History"* is meaningful and effectively utilized—with good descriptions of work performed.

15. *"Job Estimates"* are continually refined and thereby reflect improving accuracy and consistency by providing accurate, as well as challenging, expectancies.

16. *"Budgetary Procedures"* and results indicate effective control—with Production involvement in regard to control of volume variances.

17. There is an ongoing and effective *"Maintenance Skills Training Program"* for the development of individual team members; capturing their full potential and contribution, providing a sense of job fulfillment and accomplishment for each individual.

18. The *"Production Process, Maintenance and Quality Improvement Processes"* are "ONE" and embody Team concepts. Quality maintenance service in support of operational needs is evident.

19. *"Trend Charts of Progress Relative to the Goals and Targets"* are charted and displayed. Associated levels and points-in-time from which progress toward plotted Goals and Targets is monitored and plotted. These charts include trends of:
    a. *Resource Distribution/Consumption*
       i. The three forms of Maintenance
       ii. Maintenance vs. Nonmaintenance
       iii. Direct vs. Indirect Activities
    b. "Weekly Schedule Compliance, Performance and Effectiveness"
    c. "PM Schedule Compliance"
    d. "Crew Efficiency" showing that it is steadily improving.

20. There is continual search for and application of "Advancing and Applicable Technology."

11

Planning and Scheduling Fundamentals—Self-Test

11.1 APPROACH

The purpose of this chapter is first to determine your comprehension and retention of the material in Chapters 1 through 10, and second to provide a review and summary of that same material. The answers to all the questions and exercises in this chapter are contained in Appendix E. Since the material, the questions and the answers are all contained between the front and back covers of this text, you, the reader, are effectively empowered to approach this test in any manner you choose. However, to gain the most beneficial rewards from this test, the following approach is recommended:

1. Utilize a separate sheet of paper to enter the number of each question together with your response. (*Entering your answers in the book itself will preclude effective use of the self-test by anyone else.*)
2. Proceed to answer the questions and complete the exercises in this chapter based on memory only, but if unsure of the correct answer, do not guess. Leave that response blank.
3. When you have completed the exercises and answered all questions that you were able to, go back to your first blank response and search the text for the information needed to respond correctly. Continue until your answer sheet has no more blanks.
4. Now, proceed to Appendix E to score your answer sheet. If any responses were incorrect, search the text again to find the information to substantiate the correct response given in Appendix E. (*Taking at face value the Appendix E answers to questions that you responded to*

incorrectly will not etch the information onto your brain cells nearly as deep as going back into the text will.)

This four-step approach to the use of this self-test is (as assured by a leading memory/knowledge expert) the most effective method for both the comprehension of the material in this text as well as the long-term retention of the material. Good luck in your Maintenance Planning and Scheduling initiative. You are about to save your plant/company a pocketful of money!

11.2 SELF-TEST QUESTION AND ANSWER SECTION

1. Choose the correct ending to: "Maintenance is managed by managing ......"
   (a) Resources
   (b) Backlog
   (c) Equipment Criticality
   (d) Job Priority.

2. The RIME Index
   (a) Has nothing to do with maintenance; it is used in describing poetry
   (b) Compares Equipment Criticality and Job Priority to make maintenance planning and scheduling decisions
   (c) Compares Equipment Criticality and Work Class to make maintenance planning and scheduling decisions
   (d) Is used to determine maintenance trade pay grades.

3. Work Backlog is defined as:
   (a) Uncompleted work
   (b) Available work divided by unavailable work
   (c) Available work plus unavailable work
   (d) Both (a) and (c)

4. Name the three basic WO formats
   (a) _____
   (b) _____
   (c) _____

5. In order to make effective use of the planner's experience, during any available spare time the planner has, he should be assigned work supervision duties.
   True ☐   False ☐

6. An important attribute to consider for lowering the priority of—or canceling—a work order is:
   (a) Man hours to accomplish
   (b) WO age
   (c) Production facilitation of equipment access availability
   (d) Both (b) and (c)

7. What is the normal range of size for maintaining ready-to-schedule backlog? _____.

8. The maintenance _____ (*employee title*) must submit a weekly "Available Resources" report to the maintenance _____ (*employee title*).

9. Which of the following work order characteristics go into determining the work order RIME Index priority?
   (a) Work Class
   (b) Work Category
   (c) Equipment Criticality
   (d) All of the above
   (e) Items (b) and (c) only.

10. List four items of information that are required as a minimum to be included in the Weekly Backlog Status Report.
    (a) _____
    (b) _____
    (c) _____
    (d) _____

11. When ready-to-schedule backlog work is consistently less than two weeks, and as a trend it continues to get smaller, which of the following conditions is the main contributor?
    (a) The Planner is mismanaging backlog
    (b) Work is not being identified properly
    (c) The Maintenance Supervisor is submitting faulty resource reports
    (d) Production equipment is exhibiting high reliability.

12. The Maintenance Planner may automatically lower a work order's priority if the RIME Index indicates it has been set too high.
    True ☐   False ☐

13. The primary determinant for sequencing work-planning efforts is:
    (a) Job Priority
    (b) Equipment Criticality
    (c) Work Type
    (d) Items (a) and (b) only
    (e) Items (a), (b) and (c).

14. Standing work order lists generated by CMMS as due each week are not listed on the Maintenance Weekly Schedules.
    True ☐   False ☐

15. As a control document, the work order has three primary functions, which are:
    (a) ─────────────
    (b) ─────────────
    (c) ─────────────

16. Because Emergency and Urgent priority work orders are not generally "planned" by the Maintenance Planner, they are not routed to him via the WO workflow scheme.
    True ☐   False ☐

17. Regardless of whether a job is likely to be a recurring one or strictly a one-time effort, jobs should be planned to the same level of detail.
    True ☐   False ☐

18. List the six work/job package development phases of long and short-range job planning.
    (a) ───────────   (d) ───────────
    (b) ───────────   (e) ───────────
    (c) ───────────   (f) ───────────

19. What are the 9 elements of a complex or advanced work package?
    ───────────   ───────────
    ───────────   ───────────
    ───────────   ───────────
    ───────────   ───────────
    ───────────

20. Standing work orders (SWOs) are never charged to specific equipment.
    True ☐   False ☐

21. Who is the person with ultimate accountability for routine maintenance performed by the empowered Equipment Management (EM) teams assigned to Production Supervisors? _____

22. What are the three basic maintenance "responses" in the "Lean—TPM" Maintenance Organization?

    _____

    _____

    _____

23. What are the three basic maintenance "functions" in the "Lean—TPM" Maintenance Organization?

    _____

    _____

    _____

24. How is Labor Efficiency calculated?

    _____

25. What is "Resource Level Scheduling?"

    _____

26. Why is the Labor Utilization performance measure maximum value limited to around 80%?

    _____

11.3 SELF-TEST WORK EXERCISE SECTION

Note:

Work exercises in this section are basically "small-scale" samples of the types of activities that Maintenance Planners and Schedulers will be performing on the job. Although they are small in scale, the attempt has been made to incorporate all the pertinent elements of each activity. Additionally, exercises of this nature can possibly have more than one correct approach. Appendix E provides only one solution for each exercise, therefore if your solution differs from Appendix E compare yours carefully with the solution provided, ensuring that yours has met all of the criteria of a correct solution. Every attempt has been made to avoid any ambiguity in these exercises; however, if you encounter any unclear, incomplete or erroneous definitions in them, your feedback to the author(s) is encouraged.

Exercise Number 3

CPM Scheduling

As the shutdown scheduler, you have a project to schedule with 8 definable activities.
The activities (A through H) have durations as follows:

Activity	Duration
A	4
B	3
C	6
D	3
E	8
F	4
G	5
H	6

The Sequence Table for this project is shown below:

Activity	Prior	Sequence	Lead/Lag
B	A	FS	0
C	A	FS	0
D	A	FS	0
D	C	FF	2
F	C	FS	0
F	D	FS	0
F	E	FS	0
G	F	FF	0
H	E	FS	0
H	F	FS	0

Construct a CPM (AON) network schedule for the project using the node definition below.
Fill-in all node blocks for your network schedule and show the Critical Path:

Activity		
ES	D	LS
EF	f	LF

Appendix A
Job/Position Descriptions

MAINTENANCE DEPARTMENT POSITION DESCRIPTIONS

This appendix begins with a position description (PD) of the combined Planner/Scheduler function. The intent is twofold. First, obviously, is to provide maintenance management with a position description document, which can be used to update an existing PD, or create a new PD if the Planner/Scheduler function is being newly implemented. The second purpose is to provide a type of checklist of planner and scheduler duties and responsibilities, just in case you should ever wonder just what it is that you (or your planners and schedulers) are supposed to be doing. This dual intent applies to the other PDs also contained in this appendix.

The P/S Position Description:

Title: Maintenance Planner/Scheduler
Reports to: Maintenance Manager
Position Scope: The primary role of the Maintenance Planner/Scheduler is to improve work force productivity and work quality by eliminating, in advance, potential delays and obstacles through proper planning and coordination of parts, machine time and labor resources.

The Planner/Scheduler is responsible for the planning and scheduling of all maintenance work performed in the area to which he/she is assigned. He/she maintains liaison and coordination between the operations and maintenance departments; maintains appropriate records and files to permit meaningful analysis and reporting of results of completed work.

Responsibilities and Duties: In the performance of his/her duties, the Planner/Scheduler:

1. Is the principal contact and liaison person between the maintenance department and the plant departments served by maintenance and ensures that the operations areas or functions to which he/she is assigned receive prompt, efficient and quality service from the maintenance function and ensures the maintenance function is given the opportunity to provide this service.
2. The planner receives all work orders from the requesting departments of the areas to which he/she is assigned – except those for emergency work, which are requested of the appropriate maintenance supervisor for immediate attention.
3. Reviews and screens each Work Order to see that is has been properly filled out:
   a. Work scope clearly described;
   b. Check if the priority and requested completion date are realistic and provide practical lead-time;
   c. Charge numbers and other coding are complete and accurate;
   d. Authorization is proper;
   e. Discusses the details with the originating department as appropriate.
4. Makes any additional sketches diagrams, etc. necessary to clarify the intent of the Work Order.
5. Ensures that work requested is needed. If need is questioned and not readily resolved with operations or requesting personnel, refers the work order to the maintenance manager.
6. Reviews with engineering those Work Orders requiring engineering design. (Maintenance work orders which can be planned out but which require participation by shop or functional crews are copied (cross-order) and provided to the appropriate planner for planning of the supplemental work.)
7. Examines jobs to be done and determines best way to accomplish the work. Consults with requester or maintenance supervisor when necessary.
8. Obtains blueprints, drawings, instructional manuals and special procedures, as needed, from files or other sources. Makes any additional sketches, diagrams, etc., necessary to clarify the intent of the Work Order.

9. Identifies and obtains (requisitions, orders, kits as appropriate and in keeping with procedures) determinable materials, entering material needs on the Work Order. Determines if critical items are in stock by verifying availability with stores.
10. Ensures the safety needs are given a top priority in work planning.
11. Estimates jobs showing sequence of steps, the number of technicians and required labor-hours for each step. Lists determinable materials and any special tools or equipment needed.
12. Estimates cost of each work order in terms of direct labor, materials required and total cost.
13. Maintains backlog files of work orders waiting scheduling in accordance with their priority limits with an estimated completion date. Those unplanned, requiring engineering, waiting for materials, waiting for downtime, etc., are filed accordingly. When ready for scheduling, work orders are filed by supervisor by required completion day. Ideally, this filing is accomplished within a computerized system.
14. Once a job is planned and estimated, prior to scheduling, verifies the availability of parts, materials and special tools required for its execution.
15. Planner should have complete knowledge of each department's PM workload in order to better schedule work orders in that area.
16. Reviews the estimate of the current day's schedule status and forecast of labor availability on a daily basis.
17. Develops a maintenance work schedule for the maintenance area supervisor. From the backlog files for each crew, selects a group of Work Orders with labor requirements matching the capability of the identified work forces, taking into account any work carry over from jobs previously scheduled. Identifies any known skill requirements and makes such arrangements with the other planners.
18. Allocates labor and coordinates these requirements through maintenance supervisors and the maintenance manager.
19. In the selection of jobs for scheduling, meeting the deadlines established by the requesting department and maintaining preventive maintenance schedules is essential. If any Work Orders cannot be scheduled within the prescribed priority lead time, the requesting department management and the requester are promptly notified so that appropriate action can be taken to get the work done in a satisfactory and timely manner.
20. Attends meetings with the operations planning department and participates in the overall plant scheduling of the following week's work, and negotiates for downtime "windows" during which preventive and corrective maintenance requiring downtime can be performed.

Finalizes own schedules for which he/she is responsible, ensuring that the work scheduled balances with the labor-hours available so that a full day's work is provided each person.

21. Recommends equipment to be included in preventive maintenance programs.

22. Plans and schedules preventive maintenance work in coordination with operations and maintenance supervisors.

23. On the basis of firm work schedules, coordinates requisition of all predetermined parts, materials and special tools and ensures that equipment to be worked on will be available and ready. Arranges for any safety inspection, fire and standby watch.

24. Issues approved schedules together with relevant work orders and other planning documents to area supervisors. Discusses "planning packages" as necessary with special instructions or considerations to be observed in the execution of the jobs and reviews new jobs coming up in the future. (All work orders, including emergencies for tracking purposes only, come through the planner.)

25. Follows up to ensure the completed schedules and work orders are returned at the proper time. Carefully reviews completed schedules and corresponding work orders submitted by the maintenance area supervisors and monitors work order progress for any reports necessary.

26. In accordance with standard practice, provides the clerks with all documents for reporting or closeout.

27. At all times, keeps advised of the status of standing work orders, of minor maintenance and of logbook jobs.

28. Promotes the conservation of energy.

29. Maintains close contact with the other planners to ensure coordination of complex multiskill field and shop jobs.

30. Schedules weekly meetings with the operations supervisors and the maintenance supervisors concerned with the areas for which he/she is responsible, consulting them regarding facilities or equipment to be maintained. Makes recommendations to operations concerning long-range maintenance needs and, in collaboration with production, prepares a weekly forecast of all jobs expected to be scheduled for the following week. Where justified, solicits Work Orders for relatively minor corrective maintenance in order to avoid major repairs at a later date.

31. Maintains an open package of schedulable work orders, which require equipment to be shut down so that some or all of them can be performed in the event of an unscheduled shutdown of that equipment. These unscheduled shutdown lists are prepared, reviewed and updated weekly.

32. Develops a file of standard work orders (plans) for regularly recurring repair jobs, based on historical experience, to simplify the planning process.
33. Reviews with the maintenance area supervisors the actual labor expended versus estimated labor and material used for completed jobs, in order to determine corrective measures needed to improve the accuracy of estimating and improving methods of doing work.
34. Assists maintenance and operations management in periodically analyzing costs and, where necessary, recommends corrective action needed to reduce maintenance costs.
35. Keeps the maintenance manager properly informed on all abnormal or critical situations and seeks advice on matters outside of the planner's knowledge or authority.
36. Make recommendations for system improvement.
37. Maintains necessary records/files and prepares/distributes meaningful and accurate control reports.
38. Performs other tasks and special assignments as requested by the Maintenance Manager.

Position Goals: The Planner / Scheduler has the following primary goals:

1. To ensure the operations areas or functions that he serves receive the prompt, efficient and quality service from the maintenance function they need in order to operate at a high level of efficiency.
2. To ensure the maintenance function is given the opportunity to provide operations with the service it requires.
3. Accurately define and estimate work requests.
4. Properly prepare and distribute meaningful control reports.

Relationships: The Maintenance Planner

1. reports to the Maintenance Manager;
2. works closely with operations supervisors;
3. works closely with maintenance supervisors;
4. works closely with stores and purchasing personnel;
5. maintains good working relationships with all other organizational units in the plant.

Requirements, Qualifications and Selection Criteria:
The Maintenance Planner / Scheduler must:

1. have a mechanical/electrical background (necessary) and technical school background (desired);
2. have adequate trade knowledge to estimate labor-hours and materials and to visualize the job to be performed;

3. should have good oral/written communication skills;
4. must possess a demonstrated high degree of tact in dealing with both seniors and subordinates;
5. possess higher than average administrative and mathematical skills, together with a willingness and ability to organize and process paperwork;
6. have, or is able to acquire, a working knowledge of personal computers in a reasonable training period (typing skills helpful);
7. have well developed planning and organizational skills;
8. possess the ability to understand what constitutes good instructions;
9. possess the ability to read blueprints and shop drawings;
10. have an acceptable ability to produce easily understood sketches;
11. possess a thorough understanding of the proper use of work orders, priorities, scheduling techniques, etc.
12. possess the ability to keep multiple jobs in controlled motion simultaneously;
13. possess the ability to bring about order—from chaos;
14. have a mind-set and commitment to customer service;
15. possess a personal style and level of capability that will command respect within both maintenance and operating organizations.

Appendix B

Benchmarking and Best Practices

B.1. BENCHMARKING

Benchmark Definitions:

Noun

a: a point of reference from which measurements may be made; **b**: something that serves as a standard by which others may be measured or judged; **c**: a standardized problem or test that serves as a basis for evaluation or comparison.

Verb

a: to study (as a competitor's product or business practices) in order to improve the performance of one's own company.

Source: *Merriam-Webster's 11th Collegiate Dictionary.*

The benchmarking method is very much based on performance measurement. Independent performance measures are chosen before the study and the best way of achieving this performance is analyzed through investigating companies that are achieving high levels of performance in these areas. Best practice is then identified based on the performance differentials. Benchmarking is investigative as, in most cases, it has no hypothesis of which practices are "best." It analyses performance based on specified performance measures and investigates the reasons for performance differentials. These differentials typically result in the identification of best practices. This form of study is not "blind" to the effects of other practices, within the area under investigation, on performance. In this sense, the study is creating new knowledge of best practices.

The disadvantage of this type of study is that the results are very general. It is useful in terms of indicating which practices seem to work, but it does not indicate the sequence in which these practices should be implemented, the time scale for impact to occur, implementation aspects of practices, and, in most cases, the extent to which practices impact the desired performance measures. It also focuses on companies in similar industries to facilitate ease of comparison.

In an attempt to define the North American Manufacturing Industry "Best Maintenance Practices," a Benchmarking Survey of "Assessments of Maintenance Practices" of more than 200 manufacturing companies was performed by Life Cycle Engineering, Inc., a company specializing in Maintenance Engineering. The assessments evaluated the companies' practices in 21 separate, but interrelated, areas..

Following in Table B-1, are the compiled data from the benchmarking survey of maintenance effectiveness assessments.

Doing some quick comparative analysis of the data in Table B-1, it is clear from the average scores that maintenance organizations do relatively well in the elements of *Organizational Structure*, *Supervision*, and providing craftsmen with adequate *Facilities and Equipment*. They also do well at *Materials Management*. These are all elements that do not necessarily require a well-integrated process. The average scores are much lower in the elements of *Status Assessment*, *Master Planning*, *Work/Job Planning*, *Work Measurement* and *Scheduling and Coordination*. These are elements that require a more sophisticated, well-designed process with the appropriate level of discipline to follow it.

B.2 BEST MAINTENANCE PRACTICE STANDARDS

The previous section on Benchmarking did indeed refer to practices – how the art of maintenance is practiced. In order to differentiate that topic from Best Maintenance Practice (BMP) Standards, it is necessary to recognize that standards are neither arbitrary nor variable. The BMP Standards are established values of performance. You must measure your company's performance in order to compare them against the standards. Additionally, you must repeatedly measure your performance if you are to gauge your improvement. In addition to trending improvement, your performance measures are indicators of which policies are successful and which are not.

Table B-1
Maintenance Effectiveness Assessment Benchmarks

Scores	Governing Principles	Status Assessment	Objectives	Master Plan	Budgetary Control	Management Control	Organization
Lowest	0.000	0.000	0.040	0.000	0.100	0.033	0.160
Average	0.468	0.273	0.388	0.279	0.526	0.471	0.614
Highest	0.925	0.900	0.880	0.960	1.000	0.900	0.950
Median	0.475	0.200	0.360	0.160	0.500	0.433	0.580

Scores	Training	Supervision	Pride & Quality	Facilities & Equipment	Work Order System	Cost Distribution	Computer Support
Lowest	0.020	0.114	0.150	0.100	0.000	0.000	0.000
Average	0.494	0.654	0.585	0.648	0.458	0.438	0.476
Highest	0.950	0.929	0.875	0.982	0.883	1.000	0.943
Median	0.440	0.657	0.600	0.673	0.467	0.450	0.500

Scores	Equipment History	Maintenance Engineering	Preventive / Predictive Maintenance	Work Planning	Work Measurement	Material Support & Control	Scheduling & Coordination
Lowest	0.000	0.067	0.050	0.006	0.000	0.130	0.000
Average	0.444	0.436	0.514	0.355	0.267	0.589	0.395
Highest	0.900	0.857	0.865	0.850	0.717	0.890	0.880
Median	0.475	0.414	0.500	0.280	0.233	0.595	0.390

For that reason, it is advisable not to institute a multitude of changes simultaneously lest you mask what works and what does not.

Following are standards, together with their metrics, that are generally accepted as Best Practice Standards. As performance indicators, many have only a partial influence from "maintenance practices," while others are completely dependent on *how* the art of maintenance is practiced and *how well* it is performed.

- Preventive/Predictive Maintenance (PM/PdM)

Application of PM/PdM	PM/PdM hours worked ÷ Total hours worked > 30%
PM/PdM *Compliance*	PM/PdM Completed ÷ PM/PdM Scheduled > 95%

- Planned & Scheduled Maintenance Work

Planned Work	Planned Work (hours) ÷ Total Work (hours) > 80%
Planning Accuracy	Number of WOs completed within esti mate (± 15%) ÷ Number WOs completed
Schedule Compliance	Scheduled Work Completed (hrs) ÷ Scheduled Work (hrs) > 85%
Labor Efficiency	Estimated Hours ÷ Actual Hours > 85%
WO Aging	Average age of Work Orders by Priority *(refer to Chapter 6)*

- Maintenance Labor (management, effectiveness and optimization [Maintenance Engineering])

WO Management	Percentage Labor Covered by WO = 100%
Resource Management	Overtime ÷ Total Time < 5%
Labor Utilization	Labor Productive Time (wrench time) ÷ Labor Paid Time > 65%
MTBF	Mean Time Between Failures ⇒ Increasing Trend
MTTR	Mean Time to Repair ⇒ Decreasing Trend

- Total Plant Performance (includes maintenance effectiveness)

Asset Availability	Hours Asset Performs Primary Function ÷ Hours Asset Scheduled to Perform Primary Function > 95%
Asset Performance	Number of units produced (per period) ÷ Number of units scheduled to be produced > 95%

Quality of Output	Number of units produced at quality standard ÷ Total number of units produced > 95%
OEE	Operational Equipment Effectiveness = Asset Availability × Asset Performance × Quality of Output > 85%

- Stores Management and Budget and Cost Control

Inventory Stockouts	Rare (e.g., less than 1 per month – dependent on total inventory size)
Inventory Accuracy	Items Cycle Counted as Correct ÷ Total Stock Items Cycle Counted > 98%
Repair Factor	Total Maintenance and Repair Costs ÷ Total Asset Replacement Cost ⇒ Decreasing Trend
Maintenance Costs	Total costs are within ± 2% of budget
CPU	Cost per Unit (produced) = Total Plant Costs ÷ Total Units Produced ⇒ Decreasing Trend
Maintenance CPU	Total Maintenance Costs ÷ Total Units Produced ⇒ Decreasing Trend

Appendix C

Forms, Worksheets and Checklists

LIST OF EXHIBITS

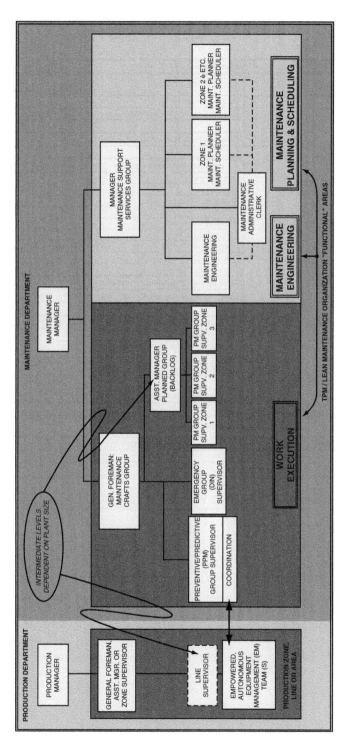

Exhibit C-1 Standard Lean Maintenance Organization

Crew/Team:				**WEEKLY RESOURCE REPORT**	
Number in Crew:					
Week Beginning:					
Available Resources Per Week					
Crew/Team Makeup	Size	ST Hrs/Wk	Tot. ST Hrs/Wk	Straight Time Labor-Hours	
Mechanics		40		Planned Overtime	
E&I		40		Man-Hrs Contracted or Borrowed	
		40			
		40		**Total Man-Hours Available Per Week**	
				Less Indirect Commitments (Weekly Averages)	
				Lunch (if paid)	
				Vacation	
				Absence	
				Training	
				Meetings	
				Committees	
				Special Assignments	
				Loaned to Other Areas	
				Other Indirect (Breaks)	
				Total Indirect Man-Hours Projected Per Week	
				Total Man-Hours Available for Direct Work Per Week	
				Direct Commitments Other Than Backlog Relief (Wkly. Avg.)	
				Planned Outages	
				Emergency/Urgent/Unscheduled (Shift)	
				Routine PPM	
				Other Fixed Assignments (Mobile Equipment)	
				Sub Total	
				Net Resources Available for Backlog Relief	

Backlog Data - Current				**Target Weeks**
Status	MH	Weeks	**Calculation**	
Available				
Total				

Exhibit C-2 Maintenance Supervisor's Weekly Resource Report Form

Maintenance Job Plan - Job Estimating Work Sheet		
Planner:	Request #: WO #:	Date: Date:

Equipment Identification:
Number   Nomenclature

Description of Work:

WorkGroup: _____   Travel :

Characteristics: _____   _____   _____   Fatigue Factor: _____   _____   _____

                    Simple   Average Complex   Light   Average   Heavy

Crew Size: _____   Skill Required: _____

Crew Balancing Required: Yes ☐   No ☐   Multi-Craft Allowance Required: Yes ☐   No ☐

Build Up of Detailed Estimates Required

Step #	Activity	Skill	Crew Size	Time Estimate
1				
2				
3				
4				
5				
6				

                                                  _____

                                        Total Direct Work (Elapsed Time)
                                            (Carry to summary below)

	Duration	Crew	MH
Direct Work Time:	_____	____	__
Job Preparation & Wrap Up:	_____	____	____
Travel Time:	_____	____	____
Total Normal Time =	_____		

x Allowance Factor of _____ = Estimated Duration _____

                                    Total _____

Estimate developed by: Slotting ☐   Detailed Estimate ☐ Historical Average ☐   Gross Estimate ☐

Exhibit C-3 Job Plan Estimating Worksheet

Weekly Backlog Report							

Date: _____ (week ending)	Planner:		Backlog Summary:				

Backlog Summary:

Priority	No. WOs	Date	No. WOs				
E/1/2	_____	_____	_____	Available	_____	_____	
3	_____	_____	_____	Unavailable	_____	_____	
4	_____	_____	_____	WO Rcvd	_____	_____	
5	_____	4 Wk. Avg.	_____	WO Cleared	_____		
Total	_____				Number	Labor Hrs.	

(Prior 3 Weeks) (Labor hours based on Available averages)

Backlog Breakdown						

Available			Age:	< 2 wks.	2 - 4 wks.	4 - 6wks.	> 6 weeks
Priority	No.WOs	Labor Hrs.					
E/1/2	_____	_____		_____	_____	_____	_____
3	_____	_____		_____	_____	_____	_____
4	_____	_____		_____	_____	_____	_____
5	_____	_____		_____	_____	_____	_____
Total	_____	_____		_____	_____	_____	_____

Unavailable			Age:	< 2 wks.	2 - 4 wks.	4 - 6wks.	> 6 weeks
Priority	No.WOs	Labor Hrs.					
E/1/2	_____	_____		_____	_____	_____	_____
3	_____	_____		_____	_____	_____	_____
4	_____	_____		_____	_____	_____	_____
5	_____	_____		_____	_____	_____	_____
Total	_____	_____		_____	_____	_____	_____

Total			Age:	< 2 wks.	2 - 4 wks.	4 - 6wks.	> 6 weeks
Priority	No.WOs	Labor Hrs.					
E/1/2	_____	_____		_____	_____	_____	_____
3	_____	_____		_____	_____	_____	_____
4	_____	_____		_____	_____	_____	_____
5	_____	_____		_____	_____	_____	_____
Total	_____	_____		_____	_____	_____	_____

Exhibit C-4 (a) Weekly Backlog Report (p. 1 of 2)

		Weekly Backlog Report		Page 2 of 2

Date: _____ (week ending)	Planner:		Unavailable Backlog Status	
Priority	**WO Number**	**Job Description**	**WO Age**	**WO Status**
___	_____	_____	_____	_____
___	_____	_____	_____	_____
___	_____	_____	_____	_____
___	_____	_____	_____	_____
___	_____	_____	_____	_____
___	_____	_____	_____	_____
___	_____	_____	_____	_____
___	_____	_____	_____	_____
___	_____	_____	_____	_____
___	_____	_____	_____	_____
___	_____	_____	_____	_____
___	_____	_____	_____	_____
___	_____	_____	_____	_____
___	_____	_____	_____	_____
___	_____	_____	_____	_____
___	_____	_____	_____	_____
___	_____	_____	_____	_____
___	_____	_____	_____	_____
		Append additional sheets as necessary		

Key: Listed in priority order by WO number

Status Codes	Meaning	Status Codes	Meaning
1	Waiting to be planned	10	Waiting for "no production scheduled"
2	Waiting for engineering/design	11	Waiting for downtime window opportunity
3	Waiting for management approval	12	Ready to be scheduled
4	Deferred - Pending Funding	13	Ready for fill-in assignment
5	Waiting for PO to be issued	14	Scheduled
6	Waiting for material receipt	15	Completed-materials/ rebuilding
7	Rcvd and ready for further planning	16	Completed-pending print revision
8	Waiting for weekend downtime	17	Closed-moved to history
9	Waiting for programmed downtime		

Exhibit C-4 (b) Weekly Backlog Report (p. 2 of 2)

Maintenance Weekly Schedule

Week Beginning:

WO Number	Job Description	Total Man hours	Mon.					Tue.					Wed.					Thu.					Fri.					Sat.					Sun.					Complete: Y or N	Comments
			Millwright	Pipefitter	Welder	Electrician	Instrumentation	Millwright	Pipefitter	Welder	Electrician	Instrumentation	Millwright	Pipefitter	Welder	Electrician	Instrumentation	Millwright	Pipefitter	Welder	Electrician	Instrumentation	Millwright	Pipefitter	Welder	Electrician	Instrumentation	Millwright	Pipefitter	Welder	Electrician	Instrumentation	Millwright	Pipefitter	Welder	Electrician	Instrumentation		

# of tasks scheduled

# of tasks completed

Tot hrs breakdown tasks

Tot available labor hours

Tot labor hrs. scheduled

% of avail labor hrs. scheduled

# of scheduled tasks not completed

# of scheduled labor hrs. not completed

% of scheduled labor hrs. completed

Exhibit C-5 Maintenance Weekly Schedule Form

CRITICALITY ANALYSIS of HVAC SYSTEM

A Asset	B Equipment	C Description	D Location	E Function Location	F Acquisition Value	G Mission Impact	H Safety Impact	I Environmental Impact	J Single Point Failure	K Preventive Maintenance History	L Corrective Maintenance History	M Reliability	N Spares Lead Time	O Asset Replacement Value	P Planned Utilization	Q Visibility	R Criticality Rating	S % Maximum
4300301	10005137	CHILLER, CENTRIFUGAL, CENTRAVAC, C-1	100-B	A-100-B-MER1	$ 73,500	5	5	5	1	4	3	2	0	4	4	0	31	62.0
4300303	10005138	CHILLER, CENTRIFUGAL, CENTRAVAC, C-2	100-B	A-100-B-MER1	$ 73,500	5	5	5	1	4	1	1	0	4	4	0	28	56.0
4300311	10005292	HEAT EXCHANGER, PLATE, ALFALAVAL, HEX-1	100-B	A-100-B-MER1	$ 19,200	5	0	0	5	5	0	0	2	2	5	0	20	40.0
4300320	10006883	PUMP, POND WATER, PWP-1	100-B	A-100-B-MER1	$ 6,800	5	0	2	5	5	0	1	0	1	5	0	20	40.0
4300319	10007043	STRAINER, DUPLEX BASKET, CW, DBS-1	100-B	A-100-B-MER1	$ 2,675	5	0	0	4	5	0	2	1	1	5	0	18	36.0
4300344	10007017	PUMP, CONDENSER WATER, CWP-1	100-B	A-100-B-MER1	$ 7,925	5	0	0	1	5	0	0	1	1	3	0	11	22.0
4300351	10007018	PUMP, CONDENSER WATER, CWP-2	100-B	A-100-B-MER1	$ 7,925	5	0	0	1	5	0	1	0	1	3	0	12	24.0
4300352	10003796	PUMP, CHILLED WATER CHWP-1	100-B	A-100-B-MER1	$ 8,500	5	0	0	1	4	0	0	1	1	3	0	13	26.0
4300360	10003797	PUMP, CHILLED WATER CHWP-2	100-B	A-100-B-MER1	$ 8,500	5	0	0	1	4	0	1	0	1	3	0	13	26.0
4400355	10005410	AIR HANDLER, AHU-1	100-1	A-100-1-MER1	$ 12,900	5	0	0	5	4	1	2	1	2	5	0	23	46.0
4300378	10005411	AIR HANDLER, AHU-2	200-B	A-200-B-MER1	$ 10,025	5	0	0	5	4	0	0	1	2	5	0	20	40.0
4300389	10007028	AIR HANDLER, AHU-3	200-2	A-200-2-MER2	$ 11,350	5	0	0	5	4	0	1	1	2	5	0	20	40.0
4300334	10007029	AIR HANDLER, AHU-4	300-B	A-300-B-MER1	$ 9,825	5	0	0	5	4	0	1	1	3	5	0	20	40.0
4300326	10007030	BOILER, HOT WATER, PATTERSON, B-01	100-B	A-100-B-MER1	$ 41,750	5	5	5	5	3	1	1	1	3	4	0	33	66.0
4300396	10007031	PUMP, HEATING HOT WATER, HHWP-1	100-B	A-100-B-MER1	$ 6,500	5	0	0	1	5	0	1	0	0	3	0	10	20.0
4300396	10007032	PUMP, HEATING HOT WATER, HHWP-2	100-B	A-100-B-MER1	$ 6,500	5	0	0	1	5	0	1	0	0	3	0	11	22.0

EVALUATION CRITERIA

G   Mission Impact: Relative criticality of system or asset on the ability to meet mission or production demands
H   Safety: Could a failure of this asset or system result in a potential safety incident?
I   Environmental: Could a failure of this asset or system result in a potential reportable incident?
J   Single Point Failure: Relative value of asset that considers work-around and ability to by-pass failure in the short-term
K   Preventive Maintenance History: Average annual cost of preventive maintenance for each asset. Absence of effective PM reduces the reliability of assets
L   Corrective Maintenance History: Average annual cost for each asset. Level of expenditures is indicative of reliability
M   Reliability: From maintenance history, this is a relative ranking based on the number of breakdowns/corrective maintenance tasks required for the asset
N   Spare Parts Lead Time: Relative measure of time required to obtain spare parts or asset replacement should a failure occur.
O   Asset Replacement Value: Relative cost to replace the asset should total failure occur
P   Planned Utilization: Relative value of the planned asset utilization. Assets that are needed more than 75% are rated highest (5)
Q   Visibility: How important is the visibility impact to the general public and staff?
R   Criticality Rating: Relative number that is the sum of the eleven evaluation criteria.

Exhibit C-6 Equipment Criticality Assessment Worksheet (Example)

Job Scoping Worksheet

JOB SCOPING WORKSHEET

Is this equipment on the PPM/PdM schedule?   Yes ζ   No ζ

If yes, initiate failure analysis and inform PPM/PdM group ζ

VISIT JOB SITE:

Verify the FI # _____   Work Order # _____

Detailed job description _____

_____

_____

Drawings ζ   Tech Manual ζ   Photographs ζ   Eng. Specs. ζ   Other ζ _____

Disconnect Location _____

Identify ES&H requirements (permits, lockouts, protective equipment, etc.)

_____

_____

Preparation and shutdown work required   _____

_____

_____

Craft: GIM _____   ELE _____   OILER _____   INSUL _____   MACH _____

Estimated Time: Hrs _____   Hrs _____   Hrs _____   Hrs _____   Hrs _____

Other Support Requirements: _____

_____

Parts and Materials: _____

_____

_____

Tools and Equipment Needed: _____

_____

List Job Steps (Continue on Back): _____

_____

_____

_____

_____

_____

Exhibit C-7 Job Scoping Worksheet

	A	B	C	D	E
	Backlog Work		=Avg. Last 4 A	=(Prev. D)+B-A	=Prev. D - UCL
Period	Completed				Backlog
week -3	186	New Planned	Completed		Relief
week -2	186	(Available)	Work	Available	Overtime
week -1	186	Backlog	(4 Week Average)	Backlog	Requested
week 1	186		186	590	
week 2	186	220	186	624	
week 3	186	250	186	688	
week 4	186	265	186	767	
week 5	186	262	186	843	
week 6	239	220	200	824	53
week 7	203	180	204	801	17
week 8	197	192	206	796	11
week 9	186	140	206	750	
week 10	186	180	193	744	

Available Backlog Worksheet

Exhibit C-8 Backlog Worksheet (Backlog Management Aid)

Company Name, Inc.	**Maintenance Planning and Scheduling**	
Written By:	Doc. No.:	Rev. No.: <u>Original</u>
	SOP-M-_____	
Approved By:	Eff. Date:	
	January 1, 2006	Page ___ of ___

1. Purpose

1.1. To establish a standard practice for processing authorized production equipment maintenance work.

2. Scope

2.1. This procedure applies to maintenance ordered by approved Work Orders that have been prepared and submitted in accordance with Reference 3.1.

3. References

3.1. Plant SOP-CO-<u>XXXX</u> – "Operations Equipment Work Order System" Standard Operating Procedure.

4. Related Documentation

4.1. Form M100-101, Maintenance Work Plan Cover/Check Sheet.

4.2. Form M100-102, Maintenance Weekly Schedule.

4.3. Form M100-120, Maintenance Group Supervisor's Weekly Resource Report.

4.4. Etc.

5. Definitions and Responsibilities

5.1. <u>Maintenance Controller (Lead Maintenance Planner/Scheduler)</u>: A second level staff management member of the Maintenance Department that, irrespective of other duties, has the responsibility for the execution of this Standard Operating Procedure including, but not limited to the supervision, guidance, training, and performance evaluation / review of assigned Maintenance Planners and Maintenance Schedulers.

5.2. <u>Maintenance Planner</u>: A staff member of the Maintenance Department that, irrespective of other duties, has the responsibility for developing comprehensive Job Plans, in accordance with this Standard Operating Procedure, for the execution of authorized Work Orders generated in accordance with reference 3.1. The Maintenance Planner is additionally assigned responsibility for the supervision, guidance, training (coordinated through the IT Department) and performance evaluation / review of assigned Maintenance Administrative Clerk(s).

5.3. <u>Maintenance Scheduler</u>: A staff member of the Maintenance Department that, irrespective of other duties, has the responsibility for scheduling, in accordance with this Standard Operating Procedure, the performance of Job Plans generated by the Maintenance Planner.

5.4. <u>Maintenance Administrative Clerk</u>: A staff member of the Maintenance Department that, irrespective of other duties, has the responsibility for assisting the Maintenance Planners and Schedulers in the performance of their functional objectives as defined by this Standard Operating Procedure. Those responsibilities include, but are not limited to, data validation and data entry into the CMMS (EAM) in accordance with this Standard Operating Procedure.

5.5. <u>Maintenance Group Supervisor</u>: A first line, supervisory level, line member of the Maintenance Department that, irrespective of other duties, has the responsibility for assignment and reporting of group maintenance craft resources for the performance of planned and scheduled maintenance work in accordance with this Standard Operating Procedure.

6. Standard Operating Procedure Responsibilities

6.1. The *Manager, Maintenance Department* is responsible for the origination, revision, approval and enforcement of this Standard Operation Procedure.

6.2. The *Maintenance Controller (Lead Maintenance Planner/Scheduler)* is responsible, as directed by the Manager, Maintenance Department, to assist in the maintenance and execution of this Standard Operating Procedure.

7. Procedure

7.1. *A system of planning, scheduling, coordinating, executing and closing out maintenance work activities should be clearly defined based upon a plant <u>Standard Operating Procedure (SOP) which should consist of five interrelated processes applicable to each maintenance job. These processes are:</u>*

7.1.1. *Plan Maintenance Job. Identify the scope of a needed maintenance job. Produce a maintenance job plan. Determine maintenance job planning category, priority and safety concerns. Identify and procure materials, and identify other maintenance task resources. Prepare the maintenance job package.*

7.1.2. *Schedule Maintenance Job. Calculate estimated start date and project resources for the maintenance job. Schedule and commit required resources and special tools/equipment items to allow performance of all maintenance tasks within the maintenance job.*

7.1.3. *Execute Maintenance Job. Initiate and perform a maintenance job and collect job information as defined in the maintenance job package.*

7.1.4. *Perform Post-maintenance Test. Verify facilities and equipment items fulfill their design functions when returned to service after execution of a maintenance job.*

7.1.5. *Complete Maintenance Job. Perform maintenance job closeout to include completion of all documentation contained in the maintenance job package to ensure historical information is captured.*

<u>*Note*</u>: **This portion of the P&S SOP must be tailored to the local organization structure and assignment of responsibilities.**

Exhibit C-9 Maintenance Planning and Scheduling Standard Operating Procedure Example

Labor/Materials/Tools Library									

Class 08 Type 01 — Description: GROTNES RIM ROLL 600

W/O	Div.	Line	Type Work					Status	
Job									

Seq. No.	Task Sequence	Labor Estimates							
		MR* M/HR		TK* M/HR		PF* M/HR		EL* M/HR	
10	Remove Top Hydromotor	2	2.00						
20	Remove Top Roll Shaft	2	5.00	1	0.50				
30	Remove Top Front Trunion	2	2.00						
40	Remove Lower Hydromotor	2	2.00						
50	Remove Lower Roll Shaft	2	5.00	1	0.50				
60	Remove Front Trunion	2	2.00						
70	Remove Flotork	2	1.00			1	0.50		
80	Remove Main Hyd Pump	1	1.00			1	0.50		
90	Remove Small Hyd Pump	1	1.00			1	0.50		
100	Remove Main Drive Motor	2	3.00	1	1.00			1	0.80
110	Remove Rim Loader	2	2.00			1	0.80	1	0.50
120	Change Bearings In Crank	2	16.00						
130	Remove Loader Control Arm	1	1.50						
140	Remove Small Drive Motor	2	2.00					1	0.50
150	Remove Pitman Arm	1	0.60						
160	Remove Guide Roll Assem.	2	3.00			1	0.50		
170	Replace Guide Roll Assem.	2	3.00			1	0.50		
180	Replace Pitman Arm	1	0.80						
190	Replace Small Drive Motor	2	3.00					1	1.00
200	Replace Loader Control Arm	1	1.50						
210	Replace Rim Loader	2	4.00			1	0.70	1	0.80
220	Replace Main Drive Motor	2	3.00	1	1.00			1	1.00
230	Replace Small Hyd Pump	1	1.00			1	0.70		
240	Replace Main Hyd Pump	2	1.50			1	0.70		
250	Replace Rotac	2	1.50			1	1.00		
260	Replace Front Lower Trunion	2	2.00						
270	Replace Lower Roll Shaft	2	6.00	1	0.60				
280	Replace Lower Hydromotor	2	2.00			1	0.50		
290	Replace Top Front Trunion	2	2.00						
300	Replace Top Roll Shaft	2	6.00	1	0.50				
310	Replace Top Hydromotor	2	2.00			1	0.50		

SEQ.	Stk. #	Qty	Material Library	P/N
1001		1	Vickers Hydromotor	45M110A-11C-10
2001	334663	1	GR-44 Roll Shaft	SAME
2002	509055	2	Torrington Roller Bearing	HJ-12415448
2003	334616	2	Grotnes Race	A-3508-107
2004	508604	2	Timkin Cone	HM-926747
2005	508602	2	Timkin Cup	HM-926710
2006	507506	4	Garlock Klosure Seal	53X3548
2007	334646	1	Sleeve	A-3508-106
2008	508045	1	Timkin Lock Nut	TAN-124
2009	508043	1	Timkin Lock Washer	TW-24
2010	525027	1	Sier Bath Cpling 2 inch Bore	1/2 Inch Key
3001		1	Front Upper Trunion	C-3508-120
3002		2	Trunion Sleeve	A-3508-170
2801		1	Vickers Hydromotor	45M110A-11C-10

			Tools Library	
2001		1	12 inch Bearing Puller	
2002		1	Medium Bearing Heater	
2601		1	100 ft/lb Torque Wrench	

* MR = Millwright   * TK = Truck Driver (Forklift)   * PF = Pipefitter   * EL = Electrician

Exhibit C-10 Labor/Materials/Tools Library Form (Example)

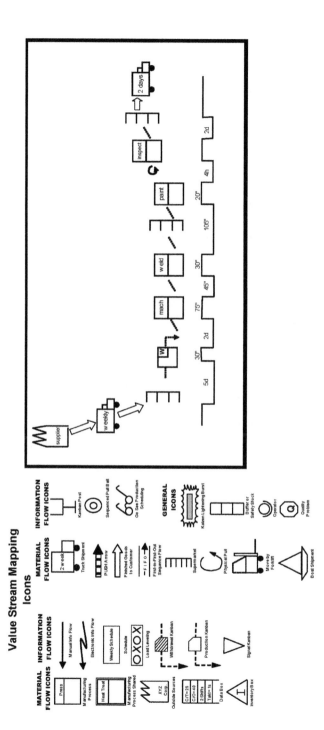

Exhibit C-11 Value Stream Mapping (VSM) Symbology and Value Stream Map Example

		Turnaround Check List for Planners/Schedulers
{	1	Determine the general scope of turnaround work from engineering schedule and from previous preventive maintenance performed during shutdown.
{	2	Determine general manpower requirements for in-house and contract labor.
{	3	Determine the general number of supervisors required.
{	4	Submit recommendations for required labor and supervisors.
{	5	Determine equipment that will be required, such as cranes, large quantities of scaffolding, compressors, welding machines or torque wrenches, etc., and if they will be available.
{	6	Determine status of materials, such as valves, internals, etc., and that they will arrive in sufficient time for checkout prior to use.
{	7	Determine pre-turnaround work for the project or other work and have work orders issued.
{	8	When the general scope of the turnaround is fairly stable, draw up a schedule for use at initial meeting to determine work force requirement.
{	9	When work force has been determined, write requests for labor. Determine workweek and request overtime authorization as required.
{	10	Write request for equipment required that will have to be rented.
{	11	Write request for labor foreman if applicable (timekeeping, etc.)
{	12	Submit personnel requisition for turnaround clerk(s) as applicable. Note: This should be done several weeks before needed to allow for approval and recruiting.
{	13	Request copy machine as applicable.
{	14	Request additional phones; one for foreman area, one for materials use and one in material trailer.
{	15	Have desks, chairs and tables moved in for coordinator, materials and zone supervisors.
{	16	Order portable toilets (early) if required.
{	17	Submit letter of request to safety for safety orientation of contractors and crews and make appointments,
{	20	See that material and tool trailers are properly supplied.
{	21	See that room and transportation arrangements are made for supervisors when there is a change. Change request submitted to senior supervisors.
{	22	Produce schedule. Distribute at turnaround meeting. Distribute final schedule as per distribution list.
{	23	See that PM work orders are produced. May have to initiate work orders.
{	24	Assemble work orders.
{	25	See that an objective for the turnaround is written and that it is given to production along with copies of all work orders.

Exhibit C-12 Planner/Scheduler Maintenance Outage Checklist

{	26	Produce readable copies of all work orders for craft supervisors to be included in packet to supervisors.
{	27	See that all work orders are activated and that all planning, including materials, is completed.
{	28	Arrange for transportation for crews as applicable (confer with craft supervisors).
{	29	Arrange with production for an area for lay-down of surplus equipment.
{	30	Arrange for an extra dumpster for waste.
{	31	Periodic update on status of preparation work and planning.
{	32	Provide a telephone and beeper list of personnel for the shutdown along with other frequently used numbers.
{	33	Secure a list of contract workers.
{	34	Provide a networked PC, printer, forms and office supplies for turnaround office.
{	35	Secure forklift if required.

Exhibit C-12 *Continued*

Appendix D

Tutorials in Brief: Control Charts and CPM Schedules

D.1 CONSTRUCTION OF CONTROL CHARTS AND THEIR USE IN BACKLOG MANAGEMENT

Control charts are a family of visual process monitoring techniques grounded in the theory of mathematical statistics. Originally, control charts were developed for use as a quality control measure in the manufacturing environment; however, more recently their application has expanded to the early detection of trends and significant changes in a variety of applications. There are six steps in constructing and using control charts to aid in backlog management:

1. Select the characteristic (backlog) to be monitored.
2. Determine the frequency and range of input.
3. Determine the chart control band midpoint and control limits.
4. Construct the chart.
5. Collect the data, plot the data and analyze the data
6. Continue to use the chart.

For this case, the characteristic is backlog labor hours, which will be input weekly as the average weekly available backlog labor hours of the previous 4 weeks. Assuming that available backlog labor hours are to be maintained within a control band of 2 to 4 weeks, the midpoint obviously is 3 weeks, the upper control limit (UCL) is 4 weeks and the lower control limit (LCL) is 2 weeks. Construction of the chart is straightforward as long as you have defined what a week of backlog is in labor hours. Since the control chart is to monitor *available backlog*, it is appropriate to use *available labor* as the yardstick. From Table 7-4 in the text, 82% of paid labor is available. In the example here, six people are in the maintenance group of concern and each works a standard 40-hour week. Therefore, available labor in 1 week is 82% of 6 × 40 hours or 196.8 hours, further the

$$\text{Mid-point} = 3 \times 196.8 = 590.4 \text{ labor hours (round-off to 590)}$$
$$\text{UCL} = 4 \times 196.8 = 787.2 \text{ labor hours (round-off to 790)}$$
$$\text{LCL} = 2 \times 196.8 = 393.6 \text{ labor hours (round-off to 390).}$$

and the control chart axes and control band parameters are defined as shown in Figure D-1.

The next step is to collect, plot and analyze the data. Representative data for this example is drawn from the Backlog Worksheet, Exhibit C7 in Appendix C. Some additional data in the Worksheet has been derived from the following assumptions:

1. Scheduled Work is 110% of Available Labor = 216 hours (196.8 hours × 110%).
2. Completed Work (Schedule Compliance) is 86% of Scheduled Work = 186 hours.
3. New Planned (Available) Backlog is 90% of new Work Orders received each week.

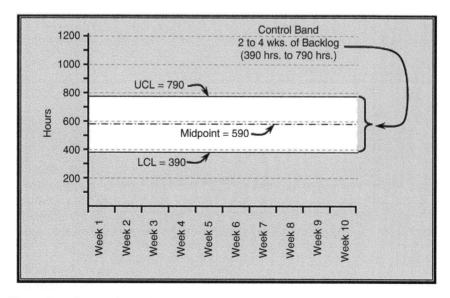

Figure D-1 Control Chart Parameters Defined

4. Backlog Relief Requested (Column E value) is the previous week's Available Backlog minus the UCL value (790).
5. Other values are calculated as shown at the top of Columns C and D (for example the Column C value of Completed Work (4-Week Average) for week 9 is the average of the Column A values of Backlog Work Completed for weeks 6 through 9).

Plotting the Exhibit C7 data on the Defined Control Chart of Figure D-1 yields the chart shown in Figure D-2. Additionally, on the chart in Figure D-2, Available Labor is represented by the vertical columns along the bottom of the chart and Backlog Relief, or overtime, is represented by the darker portions at the top of applicable columns. Finally, the Completed Work (4-Week Average) is represented by the line plot superimposed on the Available Labor columns. The resulting chart in Figure D-2 is a combined Backlog–Labor–Work Control Chart that can illustrate at a glance the state of Available Backlog management and whether or not it is under control.

Characteristic plots of various backlog conditions are illustrated in Figure D-3 (a, b and c).

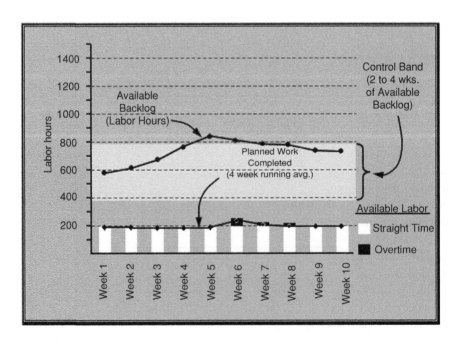

Figure D-2 Combined Backlog-Labor-Work Control Chart

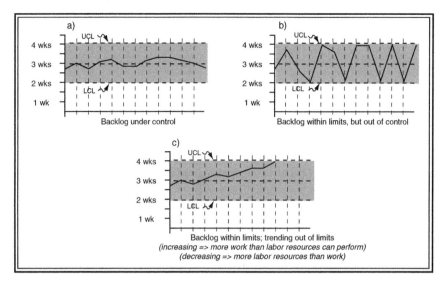

Figure D-3 Backlog Condition Control Chart Characteristics

D.2 GANTT AND CPM GRAPHICAL SCHEDULING TECHNIQUES

An activity is any significant unit of work. As you construct your schedule, you must first decide how to break the job into significant parts or activities. The time it takes for an activity to be completed, given the planned amount of material, labor and equipment, is the duration of the activity. As you define activities and then determine their durations, you may need to go back and break up activities into smaller portions. Try to break up activities based on work type and work areas. This will provide a more useable and easily understandable schedule. To get a project done within a specified time you must know when each of the activities should start. Activity start times depend on what prerequisite activities precede it. Referring to Figure D-4, look at Activity D and Activity E and assume that there are only 2 weeks to complete the project.

If the start of Activity D depends on the completion of Activity C, then the earliest that D can start is as shown on the Bar Chart. If Activity E does not depend on any other activity, then the earliest it could start is at the beginning of the job. The Bar Chart shows the latest E can start without delaying the job. The Critical Path Method determines both the *early start* and the *late start* date for each activity in the schedule. (*Bear in mind that, although the example being used here is simplistic enough to answer all the*

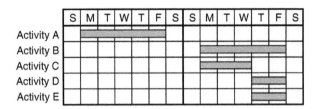

Figure D-4 Sample Project Schedule

questions posed without performing any calculations, the CPM technique described is invaluable for large, multitask jobs).

Similarly, if the completion of Activity D depends on when Activity C is completed, then the earliest that D can finish is as shown on Figure D-4. If Activity E does not depend on any other activity, then the latest it could finish is at the end of the job. Figure D-4 shows the latest E can finish without delaying the job. The Critical Path Method determines both the *early finish* and the *late finish* time for each activity in the schedule.

To make sure that a project, or phase of work, is finished on time, schedulers will often designate milestone activities or assign fixed due dates to specific activities. To finish each activity on time, you need to know what goes before and what comes after it. The bar chart does not really display this well.

In the example project above, we do not know by looking at the chart that Activity D depends upon the completion of Activity C and that Activity E could start as early as the start of the project. Bar Charts are satisfactory when everyone understands the sequence, but for large projects, it will be impossible for everyone to keep it all in mind at the same time.

One solution is to add milestone activities to the schedule. A milestone has a duration of zero (0) and will include one of two types of *plugged* dates fixing the start or end of the task. Typically, two types of plugged dates are used. The first establishes the earliest an activity can begin, called the Early Start (ES) Milestone. The second establishes the latest an activity can end, called the Late Finish (LF) Milestone (Figure D-5). When one activity must finish before another can start, the relationship between the activities is called a Finish-to-Start (FS) link.

Sequence is usually described by identifying an activity, its preceding (or prior) activity and the type of relationship that exists between the activities. Figure D – 5 can then be represented by a sequence table (Table D-1).

If activity D may start as soon as 2 days following the start of activity A. The time associated with a start-to-start (SS) sequence is called the *lead*. Another case is the situation where activity D must be finished 2 days before Activity E. The time associated with the finish-to-finish (FF) sequence is called *lag*. (Figure D-6a).

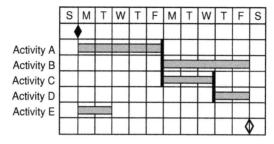

Figure D-5 Sample Project with Links and Milestones

Table D-1
Sample Project Sequence Table

Activity	Prior	Sequence
B	A	FS
C	A	FS
D	C	FS

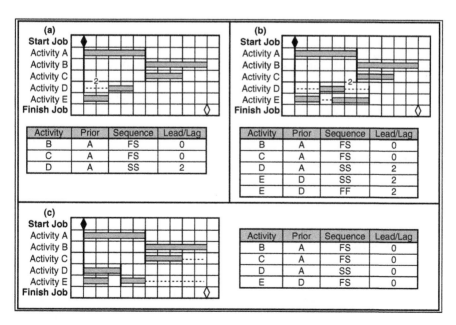

Figure D-6 (a and b) Sequence Links and Lead/Lag Relationships

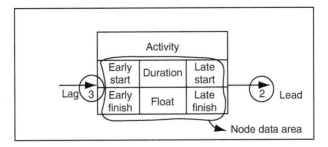

Figure D-7 CPM Node Definitions

When activities occur in parallel, they need both the Start-to-Start and the Finish-to-Finish logic to tie them together (Figure D-6b).

There will be some jobs that need to be started and finished on time to complete the overall project on time. There are other activities that may be delayed without effecting the overall duration of the project. The time between the end of activity C and the project finish, and the end of activity E and the project finish is called *float*. The float associated with, for example, activity D or E could be used by delays to activity start date or the activities taking longer than expected. Float is a property that belongs to the set of activities in the path being considered (Figure D-6c).

In order to calculate and diagram a CPM schedule, you will need to know the information contained in the sequence table as well as the duration of each activity. Figure D-7 represents the nodes that are used in the CPM diagrams and identifies the information contained within each block. Additionally, SS sequences are linked by arrows from the left side of the node *to* the left side of the node. FS sequences are linked by arrows from the right side of the node to the left side of the node, and so on. Lead and lag times are denoted by a number adjacent to the link's arrowhead. If no number is present, the lead or lag time is taken as zero.

There are three steps to calculate a CPM Schedule:

1. Forward Pass
2. Backward Pass
3. Float

Forward Pass

The first pass through the schedule (the forward pass) is begun by setting the start date. In the example provided here, for clarity, numbers will be used rather than dates with 1 representing the first day of the project and so on. The next step is to lay out the CPM nodes by following the "activity-

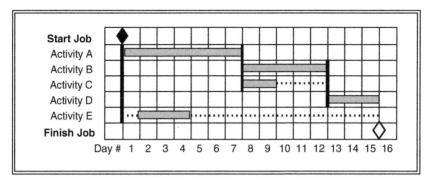

Figure D-8 Project Example Bar Chart

prior-sequence-lead/lag" data from the sequence table. Within each node, enter the known values. Figure D-8 will be used to define the CPM schedule example.

1. Add SS sequences to the first activity (A) for all those tasks that have no other priors (E). The Early Start (ES) date is the earliest date that an activity may start without being constrained by any prior tasks. It answers the question "When is the soonest you can start Activity X?"
2. Enter the remaining activities in sequence and aligned by Gantt bar alignment. You should have a diagram similar to Figure D-9.

In this example schedule, the first step is to find the Early Finish (EF) dates, where each of the question marks appears in Figure D-9. Calculate as (Early

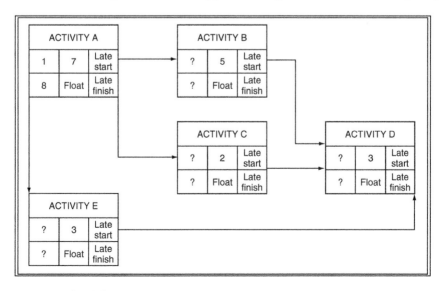

Figure D-9 CPM Graphic Layout

Start date) + (Duration) = (Early Finish date) for each activity with an Early Start date that has been set. Early Start (ES) dates for subsequent nodes are equal to the Early Finish dates for that node's prior activity. In the case where an activity has more than one prior activity, the *ES date is the maximum EF date of all the prior activities*. Therefore, ES for Activity D is 13 (from Activity B) and EF of Activity D is day 16 (ES + Duration). Next is Activity E, which is linked with a Start-to-Start (SS) sequence to Activity A. When there is no number on the linking arrow, you assume the lead is zero, therefore the *SS logic allows the linked activities to share Early Start dates (minus any lag)*, the ES for Activity E is 1 and EF is 4. Since there may be several priors for a given activity, the ES of an activity may be set by a prior activity with either a Finish-to-Start or Start-to-Start sequence. You started the Forward Pass at the first activity in the project, Activity A; the forward pass is complete when all activities have Early Start (ES) and Early Finish (EF) dates assigned to them. Your CPM graphic should now look like Figure D-10.

Backward Pass

The second (backward) pass through the schedule is begun by setting the project end date. In this example, assume that the project can finish as early as possible. When calculating a real schedule, the end date of a project will

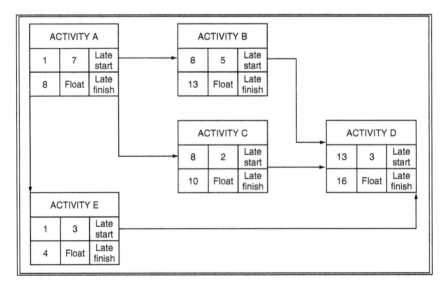

Figure D-10 Forward Pass through the CPM Schedule

be determined by a number of factors, such as Shutdown duration, other projects having yours as a prior requirement, etc. The next step is to go through the schedule from the end of the job until the beginning of the job. Consider each activity in its logical sequence. In this project example, the LF date for Activity D is day 16.

Note that when an activity fills the day square on a bar chart, it is completed at the very end of that day, which is equivalent to the very beginning of the next day.

The Late Start (LS) date is the latest that an activity can begin without delaying the completion of the whole project. The Late Start (LS) date may be calculated as *Late Finish minus Duration equals Late Start*. In the case of Activity D, LS = 13. An activity's LF date is the latest that the activity may finish without delaying any of the LS dates of the following activities. The Late Finish (LF) date is calculated as the minimum of all the subsequent activities' Late Start (LS) dates. In other words, the LF date for Activity A is the lesser of the LS dates of Activities B and C. Activity E has a Finish-to-Finish (FF) sequence with Activity D. The FF logic allows these activities to share the same Late Finish date (minus any lead).

The Finish-to-Finish sequence is only used during the backward pass calculation. You began the backward pass by setting the Late Finish (LF) date of the last activity in the project. Then you worked your way from the end of the project to the start. You have finished the backward pass when all the Late Start and Late Finish blocks have been filled in. The CPM network schedule should now look like Figure D-11.

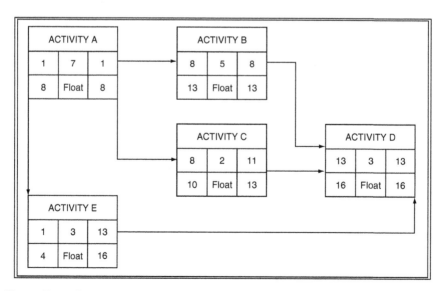

Figure D-11 Backward Pass through the CPM Schedule

Total Float

Calculating the Total Float is the last of three steps to calculating a schedule. Begin by verifying that all the dates for the early times (Early Start and Early Finish) as well as dates for late times (Late Start and Late Finish) have been calculated. Total Float (TF) is a property of a set of activities. For example, the set of Activities A and B have zero total float. This means that any delay to A or B will delay the entire project. A delay of up to 10 days is, however, acceptable for Activity E. Total float is a property of the path of activities sharing the same float value. Changing the duration of Activity C to 5 days would, for example, cause the Total Float of both Activity C and D to go to zero (0). An activity's float is calculated as Late Finish (LF) date minus Early Finish (EF) date. For example, the float of Activity D is equal to 16 minus 16 or zero.

When the float for every activity has been determined, you can find the Critical Path(s) for the project. The path through activities with minimum, or zero, total float values is the critical path, which is the sequence of activities with the least allowance for delay.

Often schedulers will talk of negative float. This common occurrence happens when the Late Finish (LF) date of an activity is earlier than the activity's Early Finish (EF). Assume that management has changed the project completion date to one day earlier, from 16 to 15. The backward pass would need to be re-calculated and total float for each activity re-computed.

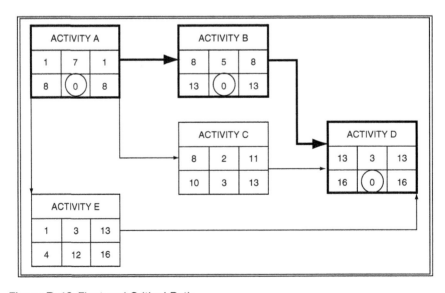

Figure D-12 Float and Critical Path

Table D-2
CPM Scheduling Operations

Forward Pass	First Activity: EF = ES − D, where ES is first day of outage Subsequent Activities: ES = EF of Prior Activity and Activity EF = ES + D Subsequent Activities with > 1 Prior Activity: ES = Maximum EF of all Priors
Backward Pass	Last Activity: LS = LF − D, where LF is last day of outage Previous Activities: LF = LS of Subsequent Activity and Activity LS = LF − D Previous Activity that is a prior to multiple subsequent activities: LF = Minimum of Subsequent LS
Float	Activity Float: f = LF − EF
Special Logic	Finish to Finish (FF) Logic »» Linked Activities share LF (minus any lead) Start to Start (SS) Logic »» Linked Activities share ES dates (minus any lag)
Critical Path	Path through network with zero (or least) total float

Negative float on an activity means that activity is behind schedule. If you were the planner/scheduler on this job, you would need to change the planned duration or sequence of any activity with negative float. The operations for constructing the CPM network schedule are summarized in Table D-2.

Appendix E

Chapter 11 Self-test:
Answers and Solutions

E.1 CHAPTER 11 SELF-TEST ANSWERS:
PLANNING AND SCHEDULING FUNDAMENTALS

1. Choose the correct ending to: "Maintenance is managed by managing . . .
   b. Backlog
2. The RIME Index
   c. Compares Equipment Criticality and Work Class to make maintenance planning and scheduling decisions.
3. Work Backlog is defined as
   d. Both a. and c. (a. Uncompleted work and c. Available work plus unavailable work)
4. Name the three basic work order formats
   a. *Formal WO*
   b. *Standing WO*
   c. *Unplanned/Unscheduled WO*
5. In order to make effective use of the planner's experience, during any available spare time the planner has, he should be assigned work supervision duties.   False ☑
6. An important attribute to consider for lowering the priority of – or canceling – a work order is . . .
   d. Both b. and c. (b. Work order age and c. Production facilitation of equipment access availability)
7. What is the normal range of size for maintaining available (ready-to-schedule) backlog? *2 to 4 weeks*

8. The maintenance *supervisor* (employee title) must submit a weekly "Available Resources" report to the maintenance *scheduler* (employee title).

9. Which of the following work order characteristics go into determining the work order priority?

    e. Items b. and c. only (b. Work Category and c. Equipment Criticality)

10. List four items of information that, as a minimum, are required to be included in the Weekly Backlog Status Report.

    a. *Ready Backlog*
    b. *Total Backlog*
    c. *PM forecasted hours*
    d. *Projected available labor hours*

11. When available backlog work is consistently less than two weeks, and as a trend, continues to get smaller, which of the following conditions is the main contributor?

    b. Work is not being identified properly

12. The Maintenance Planner may automatically lower a work order's priority if the RIME Index indicates it has been set too high.   False ☑

13. The primary determinant for sequencing work planning efforts is...

    d. Items a. and b. only (a. Job Priority and b. Equipment Criticality)

14. Standing Work Order lists generated by CMMS as due each week are not listed on the Maintenance Weekly Schedules.   False ☑

15. As a control document, the work order has three primary functions, which are . . .

    a. *Work definition and authorization*
    b. *Work planning and control*
    c. *Generate and accumulate equipment history*

16. Because Emergency and Urgent priority work orders are not generally "planned" by the Maintenance Planner, they are not routed to him via the WO workflow scheme.   False ☑

17. Regardless of whether a job is likely to be a recurring one or strictly a one-time effort, jobs should be planned to the same level of detail. False ☑

18. List the six work/job package development phases of long and short-range job planning.

    1. *Initial Job Screening*
    2. *Analysis of Job Requirements*
    3. *Job Research*
    4. *Detailed Job Planning*
    5. *Job Preparation*
    6. *Procurement*

19. What are the 9 elements of a complex or advanced work package?
    Work Order/Job Plan (includes resource requirements)
    Technical Manual (if applicable – procedures, drawings, etc.)
    Pre-Test/Pre-Maintenance Checks (if applicable)
    Bill of Materials
    Drawings/Sketches/Photographs
    Step-by-Step Procedures
    Time-By Step, By Allowance
    Re-Test Requirements/Procedures
    Post-Maintenance Notification & Reporting Requirements

20. Standing Work Orders (SWOs) are never charged to specific equipment.
    True ☐ False ☑

21. Who is the person with the ultimate accountability for routine maintenance performed by the empowered Equipment Management (EM) teams assigned to Production Supervisors?
    Maintenance Manager

22. What are the three basic maintenance "responses" in the Lean–TPM maintenance organization?
    Routine (Preventive)
    Emergency (Breakdown)
    Backlog Relief

23. What are the three basic maintenance "functions" in the Lean–TPM maintenance organization?
    Work Execution
    Planning and Scheduling
    Maintenance Engineering.

24. How is Labor Efficiency calculated?
    Labor Efficiency = (Total Estimated Hours for Work Orders Completed/Total Actual Hours for the Same Work Orders) × 100

25. What is "Resource Level Scheduling?"
    The scheduling of each labor resource for a full day's work every day that resource is available.

26. Why is the Labor Utilization performance measure maximum value limited to around 80%?
    Because (LU = Labor Productive Time ÷ Labor Paid Time) and productive time is no greater than paid time less vacation, holidays, sick time, etc.

Exercise Number 3

CPM Scheduling

As the shutdown scheduler, you have a project to schedule with 8 definable activities. The activities (A through H) have durations as follows:

Activity	Duration
A	4
B	3
C	6
D	3
E	8
F	4
G	5
H	6

The Sequence Table for this project is shown below:

Activity	Prior	Sequence	Lead/Lag
B	A	FS	0
C	A	FS	0
D	A	FS	0
D	C	FF	2
F	C	FS	0
F	D	FS	0
F	E	FS	0
G	F	FF	0
H	E	FS	0
H	F	FS	0

Construct a CPM (AON) network schedule for the project using the node definition below. Fill-in all node blocks for your network schedule and show the Critical Path

Activity		
ES	D	LS
EF	f	LF

Exercise Number 3 Solution

Appendix F
Glossary

F.1 A GLOSSARY OF MAINTENANCE TERMINOLOGY

Terminology used by Maintenance must be defined to provide a reasonable uniformity for common understanding facility-wide. A Maintenance Department, which defines its terms, utilizes them properly and informs others of their meanings, has taken a positive step in ensuring an effectively communicated Maintenance Program.

Terminology can vary from facility to facility provided it is used consistently within a specific facility or company. The following is offered as a starting point from which to develop your own glossary.

Term or Acronym	Meaning
Action Plan	The specific steps that must be taken to carry out the group's decisions, includes who does what by when.
Adjustment	Minor tune-up action requiring hand tools, no parts and less than one half hour. Adjustments restore parts or assembly relationships such as tolerance, alignment, tension and tightness.
Administrative Information	Information used to administer the Maintenance Program. Typical: Error Reports, Open Work Order Lists, etc.
Area Maintenance	A type of maintenance in which the first-line Maintenance Supervisor is responsible for all

maintenance within a reasonable-sized geographical area.

Area System
The Area System is a form of decentralization. It is the regrouping of forces into smaller, more manageable units. It is a philosophy of operation that establishes responsible units capable of solving their own problems, of running their own show, this within the framework of authority conferred, while reporting the results.

Asset
An accounting term for any physical thing owned by a plant, such as buildings, equipment, desks, software, computers etc.

Asset Replacement Value
The current accounting value of all combined physical assets in a plant.

Attitude of Error Free Work
Our personal commitment to fulfill our agreement with our customers "the first time, every time."

Authorized User
Any person who is authorized, by assignment of a password, to enter and use the CMMS / EAM.

Autonomous Maintenance
Minor, routine maintenance activities performed by production department operators as members of empowered Equipment Management (EM) teams.

Availability
(1) Informally, the time a machine or system is available for use. {Availability $= \text{MTBF} \div (\text{MTBF} + \text{MTTR})$}

(2) From the Overall Equipment Effectiveness (OEE) calculation, the actual run time of a machine or system divided by the scheduled run time. *Note that Availability differs slightly from Asset Utilization (Uptime) in that scheduled run time varies between facilities and is changed by factors such as scheduled maintenance actions, logistics or administrative delays.*

Backlog, Available
Uncompleted planned maintenance work. The total number of estimated man-hours, by trade and priority, of work required to complete all identified but incomplete planned and scheduled work. Used as an index in determining how well maintenance is keeping up

with the rate of work generation. Used also to help establish the proper size and composition of the work force. (See Open Work Order File.)

Backlog, Unavailable — Uncompleted maintenance work that has not been fully planned. Unavailable backlog is estimated to contain the volume of work, in labor hours, equal to the number of Work Orders times the average labor hours per work order in the available backlog.

Benchmarking — The continuous, systematic search for, and implementation of, better practices that lead to improved performance.

Bill of Materials (BOM) — A document listing all parts for an asset. The listing includes both stocked and nonstocked parts. The list includes the part description, manufacture, vendor, unit cost, delivery lead time and how many are required on the equipment or component.

Brainstorm — A basic problem-solving tool, which uses the unevaluated ideas of group members to generate a list of possible options. Brainstorming can generate lists of (1) problems, (2) causes, (3) solutions and (4) actions, or any list where the creativity of the group would open up new possibilities.

Breakdown Maintenance — The performance of maintenance to put failed equipment back on-line; the failure having occurred without early warning by the Preventive Maintenance or Predictive Maintenance System.

Break-In Work — Emergency or urgent work that breaks into normal work. Urgent work may have enough lead-time to be put on the Daily Schedule although it still "breaks into" the original Weekly Schedule.

Capital Funded — Work authorized by a capital fund authorization.

Category — The types of work that make up the workload performed by Maintenance. Typical: PPM, emergency, urgent and planned/scheduled (P/S) work.

Chronic Problem	One, which is characterized by long duration or frequent occurrence; one, which we have chosen to live with and have accepted as a standard.
Commonly Used Parts	Refers to a combination of standard replacement parts and hardware items that may be used on many components and pieces of different equipment.
Condition Monitoring (CdM)	A form of predictive maintenance, but distinguished by the use of installed meters, gauges and instrumentation to provide indication of equipment condition and imminent need for maintenance. Also utilizes inspection, operating tests, physical measurements, etc. Examples: a differential pressure gauge across a filter indicates the need to clean or replace the filter; an output pressure gauge on a pump, together with a pump flow meter provide data for construction of head-flow curves to determine the pump operating condition and potential need for maintenance or repair; measurement of belt tension indicates need for adjustment or replacement.
Condition Based Maintenance	Maintenance work performed based on Condition Monitoring results that indicate a predetermined condition requiring maintenance has been met.
Consensus	A group decision-making method resulting in all members agreeing to go along with a certain judgment even though one or more individuals would have handled the matter differently. The method relies on both leaders and members exploring facts, data and opinions of the membership to ensure all relevant information is considered and members have had a chance to speak their peace and voice their concerns.
Control	Control is a process by which comparisons are made between the plan and the performance, either during or after execution. Control relies on effective, complete planning and accurate quantitative observations. It is the process of comparing these two operations. In addition,

	it will be necessary to have certain dos and don'ts clearly defined for all who are going to use that structure we have called organization to obtain the desired ends of the people employed. These dos and don'ts are called policies, rules and regulations.
Construction	The creation of a new facility or the changing of the configuration or capacity of a building facility or utility. Although often performed with Maintenance Department Resources, construction work is not maintenance and should not be charged to the Maintenance Budget, where it becomes a very misleading indicator.
Coordination	Daily adjustments of maintenance actions to achieve the best short-term use of resources or to accommodate changes in operational needs. The act of synchronizing various functions and duties to obtain a desired result. It is easy to see that, if methods and efforts are not synchronized, people act at cross-purposes with each other and desired goals cannot be reached. To prevent such method failures and to ensure that all efforts shall be in the same direction, we speak of coordinating these efforts or methods.
Corrective Action	Solving problems – identifying and resolving problems; correcting a process in order to better achieve its objective.
Corrective Maintenance (CM)	Work activity to correct a problem identified by PPM activity or breakdown. It can be emergency, urgent or planned work.
Cost History	A historical picture of all cost expenditures (labor, parts, materials, etc.) against a specific unit of equipment.
Cost of Quality	The measure of what it costs when we do our job right or when we do it wrong. It measures the cost of error as well as error prevention and detection costs.
Critical Failure	A failure involving a loss of function or secondary damage that could have a direct adverse effect on operating safety, on mission, or have significant economic impact.

Critical Failure Mode	A failure mode that has significant mission, safety or maintenance effects that warrant the selection of maintenance tasks to prevent the critical failure mode from occurring.
Current State Map	Process map of existing practices. A visual method of succinctly recording the key aspects of the current structure or process in the whole or any part thereof.
Customer	The individual who receives the immediate output of your efforts (normally a coworker or boss). The "customer" is the person with whom requirements are set and agreements reached. Production is the customer of Maintenance. Lean Operations are customer focused.
Criteria	Standards on which judgments can be based. Teams use criteria to evaluate options. (See Triadic Evaluation, Priority Worksheet.)
Daily Schedule	Jobs that the maintenance schedulers/ supervisors have selected for trades' assignment that day. These jobs come from the Weekly Maintenance Schedule. Emergency and urgent work that breaks into the normal work for the week will be added as break-in work.
Deferred Maintenance	Maintenance, which can be postponed to some future date without further deterioration of equipment.
Downtime	Time when a system is not producing product. Downtime includes scheduled and unscheduled downtime.
Emergency Repairs	Immediate repairs needed as a result of failure or stoppage of critical equipment during a scheduled operating period. Imminent danger to personnel and extensive further equipment damage as well as substantial production loss will result if equipment is not repaired immediately. Scheduled work must be interrupted and overtime, if needed, would be authorized in order to perform emergency repairs. Any repair that must be done immediately to avoid injury to personnel, further damage to equipment and avoidance of costly downtime or loss of the use

of the plant/equipment. Emergency repairs should be completed within the same day of discovery.

Emergency Spares
Replacement equipment/major assemblies kept in reserve in anticipation of outages caused by man-made or natural disasters. Maintenance crews from the Emergency (Do-it-Now) Group or Facility Maintenance Division normally install this equipment.

Equipment Audit
Inspection of mechanical and electrical components of equipment to detect out-of-specifications conditions in order to assess the adequacy of shop PM and general maintenance practices. (This activity is frequently referred to as PM in Burlington plants.)

Equipment History
The chronological listing of maintenance activities, PdM/CdM results, other repair actions and equipment modifications performed on production equipment. Root Cause Failure Analysis (RCFA), when performed, is also included in Equipment History so that chronic or persistent problem identification and correction actions are archived. Equipment History is generated from Work Order closeout process. Historic repair actions also help guide current repairs. Used as the basis for developing a forecast. (See Forecasting.)

Equipment Modification
The changing of an existing unit of equipment from original design specification. Modifications may be major as when a completely redesigned assembly replaces the original, or they may be minor as when a single repair/replacement part has been upgraded or otherwise modified. Both major and minor modifications result in changes to the Bill of Material and therefore require entry into equipment history.

Expensed
Maintenance work, which does not meet the criteria for capitalization and is charged against the maintenance budget or nonmaintenance work charged to the operating budget.

Failure Effect
The consequences of failure.

Failure Mode	The manner of failure. For example, the motor stops is the failure – the reason the motor failed was the motor bearing seized which is the failure mode.
Failure Modes and Effects Analysis (FMEA)	Analysis used to determine which parts fail, why they usually fail and what effect their failure has on the systems in total.
Failure Coding	An indexing of the causes of equipment failure on which corrective actions can be based, e.g., lack of lubrication, operator abuse, material fatigue, etc.
Failure Rate (FR) or (λ)	The mean number of failures in a given time. Often "assumed" to be: $\lambda = (MTBF)^{-1}$
Five Ss	Japanese words for the five activities for improving the work place environment:

1. Seiketsu – Sort (remove unnecessary items)
2. Seiri – Straighten (organize)
3. Seiso – Scrub (clean everything)
4. Seiton – Standardize (standard routine to sort, straighten and scrub)
5. Shitsuke – Spread (expand the process to other areas)

Forecasting	The long-term projection of the best time to carry out major maintenance actions. Repair history provides a major source of these projections.
Free (Self-Service) Stock	Commonly used parts and maintenance supplies kept in Maintenance Shop, and near high maintenance areas or outside the storeroom. Withdrawal of this stock requires no requisition or other paperwork; levels are monitored by storeroom personnel weekly for replenishment requirement.
Functional Maintenance	A type of maintenance in which the first-line maintenance supervisor (or designated "expert") is responsible for direct (on-site) supervision and oversight of a specific kind of maintenance, e.g., pump maintenance for the entire plant or all bearing replacements.
Function Work	Work that does not lend itself to the area-type supervision, either because it requires specialized skills or because the nature of the work

requires maximum mobility. Functional work is performed generally on a plant-wide basis rather than by area. Examples of functional work are: electrical and instrument repairs, trash pickup, road repairs, lawn mowing and plant beautification, etc.

Future State Map
Value stream map – or other process map – of an improved process (non-value adding activities removed or minimized). A process flow that describes how one would like the process to operate in the future.

Gantt Chart
A bar chart graphically displaying activities and their durations, milestones and events and precedence relationships along a timeline. Commonly used for construction project management, Gantt charts are also used by the maintenance scheduler to facilitate his scheduling process.

Gap Analysis
An analysis of the difference between required performance and actual or measured performance; as in skills analysis and job task analysis, where gap analysis defines the training required to improve skills sufficiently to perform the job. Also used to define actions required to improve existing practices to the level of Best Practices.

Goal
The end towards which effort is directed. Goals are the steps directed towards the attainment of an objective; as such, goals are more specific than objectives; goals are milestones achieved while proceeding to objective completion.

Hardware Items
Bolts, nuts, washers, cotter pings and other items that are low in unit costs, carried in ample quantity, are readily available from suppliers and should be stocked in ample quantities for users.

Impact
The effect in dollars of a problem. High-impact problems are addressed first. Measured by analysis of "cost of quality."

Impact Analysis
The tool associated with the principle of "Measure by the Cost of Quality." This tool puts into quantifiable and nonquantifiable

	terms, the effects that arise from a problem that currently exists. A person or group would use this tool to get management's attention, prioritize problems or justify the cost of a solution.
Indirect Charges	Man-hours distributed to indirect accounting codes for nonwork activities such as safety meetings, union meetings, lunch, major delays, etc.
Infant mortality	A failure characteristic defined by component failures occurring during early life of component (1–2 months).
Inspection (PPM)	The checking of equipment with the primary purpose of determining equipment suitability for continued operation, reliability or repair needs and their relative urgency. Often accompanied by cleaning, adjusting and minor component replacement.
Inspection Route	Documented instructions for condition monitoring tasks, sorted in an effective walking path through an area.
Insurance Parts	Parts used in critical equipment and equipment components. Usage is unpredictable since the mean time between failures requiring their use is unpredictable. Their costs range from a few cents to millions of dollars. Normally, they are carried in inventory under tightest control. Not having these "insurance parts" in stock can result in extended downtime and major production losses.
Insurance Spares	Spares used on critical equipment and components of equipment. Not having these "insurance spares" in stock can result in extended downtime of critical equipment and major production losses.
JIT	Just in Time. Receiving parts, material or product precisely at the time it is needed. Avoids inventory pile-up.
Kaizen	Japanese word for the philosophy of continual improvement, that every process can and should be continually evaluated and improved in terms of time required, resources used, resultant quality and other aspects relevant to the process.

Kaizen Event	Often referred to as Kaizen Blitz – A fast turnaround (one week or less) application of Kaizen "improvement" tools to realize quick results.
Labor Effectiveness (LE)	A key performance indicator (KPI), LE = LU × LP × LQ or Labor Utilization × Labor Performance × Labor Quality (*as a percentage*), if Labor Quality is measured. (*See glossary entry for Labor Quality*)
Labor Estimate	The estimate of hours and manpower needed to complete a job, normally documented as man-hours. Estimates are determined by various techniques.
Labor Productive Time	A performance measure used in calculating Labor Utilization, it is time spent actually performing the maintenance task – often referred to as "wrench time." Labor Productive Time does not include time consumed by:
	Waiting on parts or locating parts/parts information
	Waiting on other asset information such as procedures, drawings, technical manuals, etc.
	Waiting for the equipment to be shut down
	Waiting for other trades to complete their portion of the work
	Any other delays due to the lack of effective planning and scheduling
	Any other delays due to "other factors."
Labor Performance (LP)	A performance indicator, LP is equal to Labor Scheduled Time ÷ Labor Used Time (*as a percentage*) as they refer to planned and scheduled work packages.
Labor Quality (LQ)	A subjective performance indicator, LQ is not directly measurable (normally) and can be very subjective as it is based on how well the maintenance stands up, in turn determined by such things as callbacks, time-to-fail (repeat), output (product?) quality, etc. Unless your plant has developed an error-proof method of accurately gauging Labor Quality, the Labor Effectiveness value should be (as practiced in most plants) calculated as: LE = LU × LP *instead of* LE = LU × LP × LQ

Labor Utilization (LU)	A performance indicator, LU is equal to Labor Productive Time (wrench time) ÷ Labor Paid Time (*as a percentage*).
Lean Enterprise	Any enterprise subscribing to the reduction of waste in all business processes.
Lean Manufacturing	The philosophy of continually reducing waste in all areas and in all forms; an English phrase coined to summarize Japanese manufacturing techniques (specifically, the Toyota Production System).
Lubrication Routes	An element of Preventive Maintenance. A check sheet, sorted in an effective walking path through an area, of all the equipment in that area. The document defines types of lubricant, number of lube points, type of fittings, when and how lubrication should be done and the oil change frequency.
Maintenance	The routine, recurring upkeep required to keep facilities and equipment in a safe, effective condition enabling it to be utilized at original design capacity and efficiency or some other level specified by management as the maintenance objective. Maintenance cost is normally an operating cost, although some projects, such as overhauls, performed with maintenance resources may be capitalized.
Maintenance Engineering	Sometimes referred to as Reliability Engineering, they are a staff element within the Maintenance Department whose purpose is aimed at ensuring that maintenance techniques are effective, that equipment is engineered for maximum maintainability, that persistent and chronic problem causes are discovered and corrective actions or modifications made. Responsible for maintenance optimization, review of the adequacy of repair materials used in maintenance; determination of critical parts and the adequacy of stock of replacement parts; monitor the skill levels of the maintenance work force; provide for skill level improvement, when required, through training, preparation of specifications for repair

	and new equipment selection; and other related similar actions.
Maintenance Material(s)	The parts and supplies used to maintain and repair plant equipment and facilities.
Maintenance Planning	Maintenance Planning is the advance preparation of selected jobs so that they can be executed in an efficient and effective manner when the job is performed at some future date. Additionally, Maintenance Planning is a process of detailed analysis to first determine and then to describe the work to be performed, by task sequence and methodology. It also provides for the identification of all required resources, including skills, crew size, man-hours, spare parts and materials, special tools and equipment. Maintenance Planning also requires development of an estimate of total cost and encompasses essential preparatory and restart efforts of both production and maintenance.
Maintenance Supplies	Commonly used support items that aid in maintaining and repairing plant equipment and facilities.
Maintenance Work	The repair and upkeep of existing equipment, facilities, buildings or areas in accordance with current design specifications to keep them in a safe, effective conditions while meeting their intended purposes.
Maintenance Work Order (MWO or WO)	A formal document for controlling planned and scheduled work.
Maintenance Work Order System	A means of communicating maintenance needs, planning, scheduling, controlling work and focusing field data to create information.
Maintenance Work Request (MWR or WR)	An informal document for requesting unscheduled or emergency work or a format for requesting all maintenance work. In the latter usage, once the MWR is approved, it becomes a MWO. In most maintenance operations today, the WO form itself is used to initiate requested work.
Major Repairs	Extensive, nonroutine, scheduled repairs, requiring deliberate shutdown of equipment, the use of repair crew possibly covering several

	elapsed shifts, significant materials, rigging and, if needed, the use of lifting equipment.
Manage by Prevention	Proactive practice of planning job activities aimed at keeping problems from occurring. Doing so provides the biggest return for the amount of resources expended. By usage, as opposed to management by reaction, the practice of waiting until failures occur and then taking action to repair the failed item.
Mean Time Between Failures (MTBF)	The mean time between failures that are repaired and returned to use.
Mean Time To Failure (MTTF)	The mean time between failures that are not repaired. (Applicable to nonrepairable items, e.g., light bulbs, transistors, . . .).
Mean Time To Repair (MTTR)	The mean time taken to repair failures of a repairable item.
Measurable	Capable of being compared to a standard; quality (meeting the requirements) is measured by whether or not the agreed requirements are met – yes or not. As criteria, it is the degree to which some measurement (money, time, units) may be traced to a problem.
Minor Repairs	Repairs usually performed by one man using hand tools, few parts and usually completed in less than one-half shift.
MRO Storeroom	Functional element of an organization where maintenance, repair and operating components, parts, materials and supplies are stored, usually in multiple locations. Supports both the production and maintenance operations.
Muda	Japanese term meaning waste. ‡ There are seven deadly wastes:

1. Overproduction => Excess production and early production.
2. Waiting – Delays – Poor balance of work
3. Transportation -Long moves, re-distributing, pick-up/put-down
4. Processing – Poor process design
5. Inventory – Too much material, excess storage space required
6. Motion – Walking to get parts, tools, etc.; lost motion due to poor equipment access

	7. Defects – Part defects, shelf life expiration, process errors, etc.
Non-Destructive Testing (NDT)	Nondestructive testing techniques intended to predict wear rate, state of deterioration or imminent equipment failure. These include tests such as (1) thickness measurements, dye penetrant tests, (2) Predictive Maintenance (PdM) techniques such as vibration analysis, oil sampling and analysis, ultrasonic testing, thermal (infrared) imaging and (3) Condition Monitoring (CdM) techniques such as operational testing, output measurements, etc. Normally used to refer to category 1 types of testing.
Non-value adding	Those activities within a company that do not directly contribute to satisfying end consumers' requirements. Useful to think of these as activities which consumers would not be happy to pay for.
Objective	Something towards which effort is directed; an aim of a series of goals or end of action. A strategic position to be attained or a purpose to be achieved. Objectives are statements of general plans towards which an organization's efforts are directed.
Oil Analysis	A technology of Preventive and Predictive Maintenance. It involves periodically obtaining an oil sample from various equipment and performing various analyses of the samples such as mass spectrometry, particle count and size, etc.
Operating Maintenance	Includes properly operating, caring for, cleaning and, in specific cases, lubricating equipment. This category of PM may also include certain inspections, tests, making minor adjustments, replacing frequently worn parts and correcting minor defects while equipments is operating. Operating personnel may be given the responsibility of performing much of this work, which is short-cycle preventive maintenance and may be done continuously or frequently,

	such as every shift, daily or more often than once a month.
Output	The information, service or products we supply to our customers.
Outage	A period set aside for major plant work including maintenance, equipment upgrade/replacement, plant expansion, etc. Normally the entire plant is shutdown for this work.
Overall equipment effectiveness (OEE)	A composite measure (KPI) of the ability of a machine or process to carry out value adding activity. OEE = % time machine available × % of maximum output achieved × % perfect output.
Overhauls	The inspection, teardown and repair of a total unit of equipment to restore it to effective operating condition in accordance with current design specifications.
P-F Interval	The amount of time (or the number of stress cycles), which elapses between the point where a potential failure (P) occurs and the point where it deteriorates into a functional failure (F). Used in determining application frequency of Predictive Maintenance (PdM) Technologies.
Pareto Analysis	Sometimes referred to as the '80:20 rule'. The tendency in many business situations for a small number of factors to account for a large proportion of events.
Performance Displays	Ratios, graphs, etc., which convey, at a glance, short-term accomplishments versus long-term trends.
Performance Indicators	A Performance Measure, or metric, is simply the measurement of a parameter of interest, such as Labor Hours Scheduled. A combination of several metrics yield Performance Indicators, which serve to highlight some condition or highlight a question that we need an answer to, such as "Scheduled Hours Completed ÷ Scheduled Hours = Schedule Compliance." Key Performance Indicators (KPIs) combine several metrics and indicators to yield an assessment of critical or key processes, for

	example Labor Effectiveness (LE) = LP × LU × LQ, where LP (Labor Performance) = Labor Scheduled Time ÷ Labor Used Time, etc.
Periodic Maintenance	Cyclic maintenance actions or component replacements carried out at known regular intervals, often based on repair history and regulated by current PM inspection results; includes inspecting, testing, partial dismantling, replacing consumables or complete equipment items, lubricating, cleaning and other work short of overhaul or renovation. This PPM requires equipment to be scheduled out of service and may be done at intermediate intervals, usually ranging from monthly to annually.
Planning	Determination of resources needed and the development of anticipated actions necessary to perform a scheduled major job. The orderly appraisal and guarantee of all prerequisites that is necessary to ensure completion of a given job at a predetermined time. It covers availability of ordered equipment, stores, materials, production, shutdowns, sketches, prints, specifications, etc.
Planned/Scheduled (P/S) Maintenance Work	Planned and scheduled maintenance is work, which, by virtue of cost, importance, extensive labor and materials required, etc., should be planned to ensure, when scheduled, that it can be completed with the least interruption to operations and the most efficient use of maintenance resources.
PPM Check Sheets	Lists tasks to be performed on a designated unit. It defines with instructions what component is to be inspected, what to look for, what limits are acceptable and a means to report the conditions found. It also notes any minor adjustments required if limits are exceeded but repairs are not necessary. Minor materials and parts are identified for possible periodic change out (filter, etc.). It also notes limits, which, if exceeded, require corrective maintenance.

Policy

A definite course or method of action selected from among alternatives and in light of given conditions to aid mangers to guide and determine present and future decisions about recurring situations or functions. Policies are broad direction of an authoritative nature laid down for the purpose of enabling all management decisions to be properly determined and adequately carried out in the successful attainment of the goals established.

Predictive
Maintenance (PdM)

The use of instruments and analysis to predict failure before it takes place, based on a change from normal conditions. Examples include vibration analyzers to detect slight increases in high-frequency vibration amplitude in bearings, infrared scanners to detect poor electrical connections. Predictive maintenance consists of three factors: (1) monitoring, (2) trending and (3) diagnostics. The resultant information allows for the application of timely corrective maintenance based on a machine's actual condition. This is part of Preventive Maintenance. All predictive work stems from documented instructions or chick sheets that are issued from documented schedules.

Prevention Analysis

The tool associated with the principle of "Manage by Prevention." By using this tool, problems can be brought out and dealt with before an activity is performed. Planning before the fact is always less expensive than reacting and patching after the fact.

Preventive/Predictive
Maintenance
(PM/PdM)

Equipment inspection and nondestructive testing to determine future repair needs, their urgency, lubrication and minor adjustments to prolong equipment life. This may include cleaning, adjusting and minor component replacement. Inspections of mechanical and electrical components of equipment are designed to detect out-of-specification conditions (excessive heat or vibration, lack of lubricant, etc.) in order to correct such conditions before premature failures results. Early warning effort to identify developing problems and need for early corrective maintenance:

1. Routine visual and audio check of equipment
2. Lubrication
3. Housekeeping and
4. Predictive Maintenance

All work credited to PM is from documented instructions or check sheets that are issued from documented schedules. Minor work is to be limited to half hour and the use of predetermined parts or materials and the use of small hand tools such as channel locks, 10-inch adjustable wrench, straight blade screwdriver and flashlight.

(Preventive Maintenance uses human senses; Predictive Maintenance applies sensory instrumentation.)

Preventive Maintenance Overhaul and Shutdown	Includes major work involving dismantling and inspecting equipment before breakdown occurs. It includes replacing or reconditioning equipment and components, which have reached or are approaching their theoretical maximum life limit determined by predictive techniques. This activity includes major overhauls and is long-cycle PPM performed generally at intervals varying from six months to more than a year. This category of maintenance does not include overhauls, the need for which is generated outside of PPM guidelines, such as at fixed intervals regardless of equipment condition.
Priority	The relative importance of a single job in relationship to other jobs, operational needs, safety, equipment condition, etc., and the time frame necessary in which the job should be done. It is used primarily for Planned Work, which subsequently will be scheduled.
Priority Worksheet	A problem-solving tool used to help in decision making, similar to *Triadic Evaluation* except that options are rated against several criteria.
Proactive Maintenance	The collection of efforts to identify, monitor and control future failure with an emphasis on the

understanding and elimination of the cause of failure. Proactive maintenance activities include the development of design specifications to incorporated maintenance lessons learned and to ensure future maintainability and supportability, the development of repair specifications to eliminate underlining causes of failure, and performing root cause failure analysis to understand why in-service systems failed.

Problem/Cause
A serious condition or situation, which prevents us from doing our job right the first time. The problem/cause is the difference between our current output and our desired output. It could be chronic or sporadic.

Problem/Cause Statements
Identified effects, clearly understood by the group and stated in terms, which reflect the impact of the problem/cause.

Procedure
A series of steps followed in a regular definite order in which activities or tasks are to be carried out.

Process Mapping
Technique for indicating flows or steps in a process using standard symbols. Used to facilitate process improvements.

Project Work
Construction, installation, equipment, relocation or modification of equipment, buildings, facilities or utilities to gain economic advantage, replace worn, damaged or obsolete equipment, satisfy a safety requirement, attain additional operating capacity or meet a basic need. Usually is capital-funded, seldom true maintenance.

Pull system
A manufacturing planning system based on communication of actual real-time needs from downstream operations ultimately final assembly or the equivalent – as opposed to a push system which schedules upstream operations according to theoretical downstream results based on a plan which may not be current.

Purchase Order
The authorization documentation for obtaining direct-charge materials or services from vendors or contractors.

Quality	Meeting the measurable requirements agreed upon with our customers. When we meet these requirements, we have "quality"; when we don't, we have an "absence of quality."
Rebuild	The repair of a component to restore it to serviceable condition in accordance with current design specifications.
Requested Maintenance	A request for maintenance service, which did not emanate from the PM system, but did provide sufficient lead-time to allow proper, proactive, planning and scheduling.
Regulations	Rules that are concerned with methods of activity or performance. Both rules and regulations are derived from the broader *policies* of the organization.
Relocate	Move fixed equipment to a different stationary location.
Reliability	The dependability constituent or dependability characteristic of design. From MIL-STC-721C: Reliability (1) The duration or probability of failure-free performance under stated conditions. (2) The probability that an item can perform its intended function for a specified interval under stated conditions.
Reliability Centered Maintenance (RCM)	The process that is used to determine the most effective approach to maintenance. It involves identifying actions that, when taken, will reduce the probability of failure and which are the most cost-effective. It seeks the optimal mix of Condition-Based Action, other Time- or Cycle-Based action, or Run-to-Failure approach.
Reliability Engineering	See Maintenance Engineering
Repair	Repair is the restoration of an asset to a condition equivalent to its original or designed capacity and efficiency by replacement of parts or after deterioration, overhaul to enable continued processing of materials.
Repair History	The chronological listing of significant repairs made on key units of equipment and the analysis of these repairs to help identify chronic, repetitive problems, failure trends and the life span of critical components.

Repetitive Maintenance	Maintenance jobs, which have a known labor and material content and occur regularly (daily, weekly, etc.).
Replacements	Replacement covers badly worn parts (chains, belts, bearings), which are no longer capable of adjustment and must be replaced. Consumable parts, such as seals and gaskets, are other examples of replacement parts. Scheduled replacement is performed to avoid costly repairs.
Reposition	Move mobile equipment to a new working location.
Requirement	Documented and agreed upon standards and objectives, this tells us what we are to do in our job, how it is done and what is expected as the output of our efforts.
Requirements Analysis	The tool associated with the principle "Meeting the Requirements." A basic tool used to analyze supplier/customer relationships to make sure clear requirements have been communicated.
Return on Investment	A measure of the cost benefits derived from an investment. ROI (in %) = [(total benefits - total costs) ÷ total costs] × 100
Root Cause (Failure) Analysis	Root Cause (Failure) Analysis – RCA or RCFA – is the maintenance engineering discipline, which directs attention to repetitive or costly failures in assets in order to determine the underlying weakness (defect), which has caused the failure. Once the nature of the defect is isolated, it is then possible to design appropriate engineering action to eliminate or minimize impact of the failure. The documentation, prioritization and analysis of the mechanism that caused the failure are an important part of the improvement effort.
	Involving maintenance personnel in the analysis phase provides opportunity to learn and increase technical knowledge, contribute significantly to breakdown prevention and provides "hands on" experience sometimes crucial in developing practical corrective solutions.

These activities are basic to satisfying higher psychological needs as defined by behaviorists. As in other participative management efforts, the hourly maintenance person must be provided the training necessary for participation. This involves training to recognize the types of failures and their causes in various machines and machine components such as bearings, gears, etc.

Another equally important benefit of worker involvement is an understanding developed concerning the failure modes, which reduce machine availability. Through failure analysis, the maintenance person learns how to recognize the symptoms and "tracks" of defects and how they occur. Such understanding is the best possible training for "quality maintenance workmanship." When the consequence of poor or sloppy maintenance procedure is understood, mechanics are much better conditioned to understand the necessity of and practice of proper method.

Routine Maintenance (repetitive work)	Janitorial work, building and grounds work. Often applied to personnel who perform highly repetitive work such as tool sharpening, etc. Services performed consistently in the same manner: includes actions such as grass cutting, freeze-protection and janitorial services.
Rules	Standards or guides for performing specific operations or limiting the activities of people.
Scheduling	Determination of the best time to perform a planned maintenance job to appreciate operational needs and the best use of maintenance resources. The process of accomplishing planned engineering and maintenance work, at a predetermined time, which coincides as close as possible to the required completion date. It implies the orderly use of engineering and trade skills to accomplish the greatest good at any particular time.
Schedule Compliance	The number of planned and scheduled jobs or PM services actually accomplished during the

	period covered by an approved schedule. The number of actual man-hours worked against scheduled man-hours (%).
Scheduled Maintenance	Extensive major repair, rebuilds, overhauls, major component change-outs, etc., requiring advanced planning, lead time to assemble materials, scheduling equipment shutdown to ensure availability of repair-facility space and allocation labor.
Shutdown	A "shutdown" is defined as the scheduled removal of a facility from service to open, clean, inspect, repair, add, alter, close and test operating components; then return of the facility to service with a predetermined interval of time.
Slotting	A method of estimating labor hours using established labor benchmarks or labor library.
Spares (Specialized Spare Parts)	Parts that are used in and are unique to specific equipment components and equipment.
SKU	see Stock-keeping Unit
Solution	An activity, which eliminates (or reduces the impact of) a root cause of not meeting our goal.
SOP	See Standard Operating Procedure.
Specifications	Technical definition of configuration or performance requirements to meet intended utilization of equipment or materials.
Sporadic Problem	One, which is characterized by only occasional occurrence or by scattered instances; also a problem, which triggers alarm signals and requires a response.
Standard Replacement Parts	Parts that can be used on more than one component or piece of equipment. Suppliers for a number of users may carry these parts in stock. Delivery lead times are predictable so stock outs can be managed.
Standing Operating Procedures (SOP)	A written procedure used to ensure reasonable uniformity each time a significant task is performed or process is executed.
Standing Work Order	A reference number used to identify routine, repetitive actions.
Stock Issue Card	The authorized document for making stock material withdrawals.

Stock keeping Unit (SKU)	An item in Stores with an assigned inventory number. The unit quantity of a specific SKI in inventory may be one or more; it may also be out of stock.
Symptom	Evidence that a problem/cause exists. This evidence needs to be clarified to determine the impact of the problem/cause.
System	A system is nothing more, or less, than an orderly, habitual or routine method, or methods, by which the regular activities of a part of a business, like maintenance, are carried on. Within Maintenance, System usually refers to the Maintenance Management Information System.
Target	A specific, quantitative measurement established in order to measure progress towards a goal.
Time Distribution Card	The authorized document for reporting the use of labor against a specific job.
Total Productive Maintenance (TPM)	A manufacturing led initiative for optimizing the effectiveness of manufacturing equipment. TPM is team-based productive maintenance and involves every level and function in the organization, from top executives to the shop floor. The goal of TPM is "profitable PM." This requires you to, not only prevent breakdowns and defects, but also to do so in ways that are efficient and economical.
Unscheduled Repairs	Unscheduled nonemergency work of short duration. Work can be accomplished within approximately one week with little danger of equipment failure in the interim period. One person typically makes these repairs in less than 2 hours with materials needed in about 50% of the instances. Unscheduled repairs should be completed within one week after they are discovered.
Value adding	Those activities within a company that directly contribute to satisfying end consumers, or those activities consumers would be happy to pay for.
Value Stream	The specific value adding activities within a process.

Value Stream Mapping	Process mapping of current state, adding value or removing waste to create future state map or ideal value stream for the process.
Verbal Orders	A means of assigning emergency work when reaction time does not permit preparation of a work order document. (Use of verbal order must be accompanied by a procedure, which ensures resource use is reported.)
Vibration Analysis	This is a part of Predictive Maintenance. It is the work performed on mechanical rotating equipment to evaluate any undesirable changes that might indicate the beginning of failure. This may lead to the recommendation of a logical course of maintenance actions to correct the problem before secondary damage or catastrophic failure can occur.
Weekly Forecast	A forecast listing of planned work from the backlog, to be scheduled during the following week.
Wiebull Distribution	A statistical representation of the probability distribution of random failures.
Work Force	The personnel who carry out the maintenance workload.
Workload	The number of man-hours required to carry out a maintenance program.
Work Sampling	The statistical measure of labor utilization to determine productivity.

Index

Page numbers followed by f and t indicate figures and tables, respectively.

Printed and bound by CPI Group (UK) Ltd, Croydon, CR0 4YY

03/10/2024

01040434-0006